Tools of Their Tools

Tools of Their Tools: Communications Technologies and American Cultural Practice

Edited by

Grzegorz Kość and Krzysztof Majer

CAMBRIDGE SCHOLARS
PUBLISHING

Tools of Their Tools: Communications Technologies and American Cultural Practice
Edited by Grzegorz Kość and Krzysztof Majer

This book first published 2009

Cambridge Scholars Publishing

12 Back Chapman Street, Newcastle upon Tyne, NE6 2XX, UK

British Library Cataloguing in Publication Data
A catalogue record for this book is available from the British Library

ISBN (10): 1-4438-0978-0, ISBN (13): 978-1-4438-0978-8

We dedicate this book to the memory of Ewa Grzeszczyk, one of our contributors, who died tragically on May 22, 2009. She was an excellent scholar and a wonderful colleague, who will be greatly missed.

TABLE OF CONTENTS

PREFACE

All the essays collected in this book explore the role of communications technologies during the last 150 years in American cultural practice. By communication technologies we mean audio and visual reproduction technologies, analogue telecommunications such as traditional telephony, radio and television broadcasts, digital telecommunications, computer-mediated communications, telegraphy, and computer networks.

The study of the impact of telecommunication technologies is a way to explore the various flows and tensions of American culture. How has American society molded communication technologies? How have they, in turn, shaped American history? Are Americans still, in the words of Thoreau, "tools of their tools"? More so or less than during the philosopher's Walden days? How do America's cultural, ethical, and economic assumptions determine and limit the ways in which telecommunications function in American society?

Fascinating questions abound. How do contemporary telecommunications transform the sense of America's timespace? To what extent have mass media failed to breed cultural and political conformity? Have the newest digital technologies been more successful in producing diversity and difference? Our contributors have investigated the rise of electronic communities—how have the new media transformed the patterns of social interactions in terms of politics and sexuality? Does the Internet and the Web foster the heterogeneous character of the U.S.? We examine the forms and significance of many-to-many communication via e-mail, Internet forums, blogs and chat rooms. What are some of the new interventions in the social construction of emerging telecommunications and the innovative uses of the new and not-so-new, analogue and digital technologies? Last but not least, how are analogue and digital telecommunications represented in literature? What may be the import of digital reading and writing—digital re-reading of old literature, hypertexts, cybertexts and e-literature? These and other related issues are explored in this book.

Andrew Ross looks at the US national and social costs of cross-border technology transfers and web-related outsourcing of high-skill jobs. Offshore, especially to China, went not only much of assembly platform work but also, right in its wake, R&D sectors. PRC's rise to the position of a superpower creates problems of its own for Americans. According to Ross, Americans have basically two weapons against processes which their multinationals have helped to unleash. One is their commitment to create another superpower on China's turf, i.e. India; another the tightening of legal regulations on the licensed use of technologies, which would assure that the technology transfers are accompanied by reverse flows of revenue. The latter efforts, Ross predicts, are likely to be frustrated by a broad Internet-networked coalition of hacker-minded professionals who gradually begin to understand they can make the blue-collars of the coming decades.

In their melancholy, kaleidoscopic essay on the history of American radio, David Jones and Joanna Waluk pay tribute to its nation-building role in the difficult decades from the 1930s to the 1960s, which were ridden with crises threatening to rip apart the fabric of American society. For balance, Raffaele Asquer outlines the limits of the airwaves' ideological impact by sampling the mixed public response to FDR's fireside chat of March 9, 1937 in defense of his rather insupportable Court-packing Bill. If not perfectly tight, the mainstream media's grip over Americans has never significantly slackened. Most of us have gotten used to the complaint that television-driven personality politics has made a mockery of real democratic debate. In his contribution, Maciej Turek turns his media critic's eye on American vice-presidency alone; he demonstrates that the marked neglect of this institution as well as of the televised "Veep" debates by the American public may be a symptom of national carelessness.

Małgorzata Gajda-Łaszewska draws our attention to a worrying alliance being forged between corporate media and PR officers. She shows how various business and government actors go out of their way to be sourced by the media by providing them with Video News Releases, news clips of their own making. This is not to say, however, that the mainstream media are only propagandistic and allow no dialogue: Aleksandra Różalska praises the popular miniseries *Sleeper Cell* as a noteworthy cultural intervention subverting the prevalent "Orientalist" bias. By creating Darwyn Al-Sayeed, an African American Muslim and an undercover FBI agent who infiltrates a terrorist outfit, the makers of *Sleeper Cell*—Różalska maintains—indicate a worthwhile direction for the American television, a step towards dismantling Western ethnocentrism as outlined by Edward W. Said.

We have witnessed too much bashing of the mainstream media not to transfer our hopes to the Internet. Collapsing space and time more successfully than any other communication technology thus far, the Web has brought together the most remote areas and the most isolated social or ethnic groups. It remains to be seen, however, whether it will deliver the long-awaited "global village," a perfectly open, fully participatory environment which Marshall McLuhan saw incipient in the success of radio and television. Mirosław Miernik assesses the accuracy of McLuhan's predictions against the media landscape some 40 years after the publication of *The Gutenberg Galaxy* and gauges the usefulness of the Canadian's notorious interpretive instrumentarium. Richard Profozich offers a comprehensive, and on the whole laudatory, review of new developments in online citizen journalism. Today from many quarters come alarmed voices that the "need to know" of responsible citizenry is being replaced by yahoo junkies' infosnacking. Without explicitly denying the validity of this argument, Profozich is not unhappy to see the media priests and punditocrats cut down to size by bloggers who, with the help of broadband and WiFi connections, can pool expertise more efficiently than any of the traditional think tanks.

In her sociologically disciplined analysis "Fan Culture in the Internet," Ewa Grzeszczyk takes notice of the rising power of online fan communities, the so-called Generation C, which has its new ways of actively influencing the media content. On a more skeptical note, Anna Krawczyk-Łaskarzewska expresses a mixed opinion of citizen film reviewing. In his well-known August 2006 *New Yorker* piece, Nicholas Lemann, a professor of the Graduate School of Journalism at Columbia, poured cold water on the enthusiasts of "journalism without journalists." Although Krawczyk-Łaskarzewska is no less questioning in her estimation of the quality of online film reviewing without professional film reviewers, she cautiously admits that online amateur journalism "offers the potential for creating a multilayered critical space."

Notwithstanding the excellence of her essay, Krawczyk-Łaskarzewska's difficulty to come to terms with an Internet-driven change in the public realm is common. Her tendency to see citizen film reviewers as mere impostors and her reluctance to embrace the newly emergent rhizomic space in which films are being assessed today may be just part of a much wider resistance against hitching human subjectivity up to the technological other. Such breakthrough has long been awaited, and called for, by Gilles Deleuze and Felix Guattari, whose enthusiasm for technological immanence is explicated in this book by Marek Wojtaszek. Wojtaszek reads *The Matrix* as an illustration of Deleuzian concepts, and as offering a radical reconceptualization of human experience into a virtual sensibility—cubistic, dynamically evolving, and always ready to make itself comfortable in ever newer realms offered by autonomously evolving technologies.

Paradoxically, it is not "techno-logic" alone that rules even cutting-edge labs and design shops. In fact, new technologies are driven by certain age-old myths as well, questioned today not only by philosophers but also by novelists and film directors. Dorota Golańska, for instance, examines two relatively recent SF movies, Steven Spielberg's *Minority Report* (2002) and *AI* (2001), to argue that they effectively challenge many of the long-established neo-positivist paradigms such as the privileging of the natural over the artificial, or the dominance of seeing in epistemological enquiry, perpetuated by the most recent developments in science and medicine (such as imaging technologies of deep body structures). They do so, even though American SF movies seem ill-suited to dispute values that, after all, lie at the foundation of the genre and national mythology.

Our contributors have also mapped a range of literary responses to the impact of communication technologies. Invoking Edward Balcerzan's concept of multimedial genres, Alicja Piechucka reinvestigates T. S. Eliot's interest in incorporating a variety of genres and techniques within one work. Piechucka goes so far as to argue that the language of T. S. Eliot and the French symbolists anticipates the media-saturated experience of our time and reminds us that some critics (notably Richard Schusterman) have described the author of *The Waste Land* as a postmodernist. Interested in more active responses, Julia Fiedorczuk looks at the fortunes of today's avant-garde poets. Wishing to avoid sterile aestheticism and engage in social praxis, they can do so only from a certain remove, safe from "the means-ends rationality of the bourgeois everyday," a position that —when all is said and done—has many features of the rejected "ivory tower." Communication technologies can hardly help escape the predicament. To begin with, they are part of the problem: nothing is more formulaic than downloadable e-motional greeting cards or media-ized political discourse. But fighting them, L=A=N=G=U=A=G=E poets discovered, is like squeezing the balloon. Charles Bernstein's innovative orthography may be subverting the ideal of transparent meaning, but involuntarily it also risks being co-opted into the mass enthusiasm for spelling glitches, acronyms and related instances of accelerating the already speedy text messages and emails. Fiedorczuk sees more hope for "virtual communities" that one day may churn out collective poems, no longer adhering to the myth of the autonomous authorial subject.

A glimpse into fictional strategies of dealing with the ascendancy of the digital is offered by Marta Koval in her analysis of John Barth's 2001 novel *Coming Soon!!!*—a characteristically autotelic dramatization of the aesthetic antagonism between printed and electronic literature ("p-fiction" and "e-fiction," respectively). Examining the ontology of the written word at the turn of the millennium, Barth's inter- and intratextual work playfully mimics hypertext and tantalizes the reader with false promises of navigational freedom. Although the writer declares his interest in hypertextuality mainly as a metaphor for language, Koval argues that various deployments of the idea of "mingling of essences" throughout *Coming Soon!!!* encourage a reading of the work as a narrative mediation between the two modes.

A broader antagonism is sketched by Kate Delaney, who invokes C. P. Snow's legendary pronouncement on the "two cultures" in her discussion of two relatively recent novels which imagine a rewarding confrontation between science and the humanities in a university setting. The novels in question are Richard Powers's *Galatea 2.2*, in many ways a reenactment of the Pygmalion myth, and *Thinks...* by David Lodge, which revisits some of the novelist's pet themes and situations. Delaney argues that both works envision a veritable symbiosis between a frustrated writer and a confident scientist: the former, forced to defend and thus to reconsider the relevance of literature, is reinvigorated in the process and finally able to deal with his or her writer's block.

Some of our contributors are also interested in the impact of virtuality on a writer's public image. Focusing on the celebrated American literary recluse Cormac McCarthy, Hanna Boguta-Marchel examines the virtual persona constructed by net-savvy readers and enthusiasts through Web sites such as cormacmccarthy.com. Boguta-Marchel also comments on the irony between the fans' reliance on technological progress and its function in McCarthy's novels: conspicuously absent from the narrative, or present only in the form of its dire consequences, technology often looms as a grave threat to the very existence of mankind.

Beata Zawadka's essay demonstrates that reflection on contemporary communication technologies may illuminate literary works which predate their arrival. In her unorthodox reading of Octavia Butler's 1979 novel *Kindred*, Zawadka sees the notion of time travel as an integrative strategy, a prescient metaphor of the World Wide Web. Apart from commenting on current cultural practice—Zawadka argues—Butler's metaphor appears to point towards an imagined, idealized future, in which empathy becomes a key to healing the trauma of slavery and tightening the bonds of the American society.

Literary approaches to traditional means of communication are also featured in the volume. Noel Polk focuses on a few mechanical devices which appear in Eudora Welty's memoir, *One Writer's Beginnings*. A prototypical Victrola and a couple of telephones are invoked in the essay to suggest that underneath the enthusiasm for the modern era, embodied by Welty's father, there lurks what the writer herself experiences as a discernible "growl," an unsettling, discordant quality. What is ultimately channeled through these mechanisms, Polk believes, is not so much human communication—a clear focal point of Welty's writing—as a variety of "one-way communiqués"; far from ascribing all the blame to technology, however, the author of the essay entertains the thought that this mechanical "noise" may in fact highlight a difficulty inherent in any contact between people.

Tomasz Basiuk traces the distortive influence of telephony on amorous relationships in two gay plays, Mart Crowley's *The Boys in the Band* and Robert Chesley's *Jerker, or The Helping Hand*. Basiuk invokes the Derridian revision of the assumed primacy of speech, applied by Lee Edelman to reveal an implicit coding of reproductive and non-reproductive sexuality, to describe ways in which the spoken word, when transmitted through the telephone, becomes distant to itself and thus approximates writing as understood by the French theoretician. The act of telephoning becomes a deconstructive procedure, subjecting the seemingly direct meaning to unforeseen, circumstantial alterations, and compounding spatial distance by emotional detachment.

An alluring intersection between technology, diasporic realities and gender is offered by Izabella Kimak, who analyzes the role of telephony in selected works by two South Asian American woman writers, Chitra Banerjee Divakaruni and Amulya Malladi. In their fiction, the telephone emerges as a double-edged, ambivalent instrument of power: used by the family in the homeland as a tool of surveillance and a means to guarantee the female émigrés' obedience to tradition, it also becomes a potentially subversive medium of sororal succor across the Pacific.

Analyzing the depictions of African Americans in the era of the "graphic revolution," Matthew Wilson argues that the invention of halftone photoengraving may have fostered the objectification of those portrayed on account of the supposed veracity associated with the new technique. Wilson cites examples from elite magazines such as *Harper's* and *Scribner's*, as well as an instance of self-depiction in the narrowly circulated *Colored American Magazine*, intended as counterbalance to the prevailing imaginary. Paradoxically—and tragically—regardless of the intent with which the illustrations were published, they seemed automatically to become incorporated into the discourse of the dominant ideology.

Also present in the volume are illuminating evaluations of technology as experienced in the performing arts. In his essay, Radosław Rybkowski examines the impact of nineteenth century technological innovation on the American theater and its audience. Rybkowski engagingly explains how such devices as the gas table, the lime lamp, the hydraulic stage elevator or the folding chair gradually eliminated or lessened problems encountered both by the purveyors and recipients of the theatrical experience, ultimately leading to changes in the patterns of artistic communication.

In her appraisal of the art of John Cassavetes, a pioneer auteur of American cinema, Elżbieta Durys maintains that the initial rejection of his work in the sixties and seventies was caused, among others, by the director's highly innovative approach to the film medium. Rejecting both the "zero degree" poetics championed by Hollywood and the achievements of foreign and domestic art film—Durys argues—Cassavetes found inspiration in the documentary style of direct cinema. A maverick filmmaker, he did not receive appropriate recognition until the nineteen-eighties, by which time audiences had learned to appreciate the technical innovation which made his works unique.

If the book begins by describing ways in which the new technologies have exacerbated nationalist anxieties in the U.S. and transmogrified the media, it closes with an inquiry into their impact on American arts. Like other cultural practices, the arts are as much on top of communications technologies today as they are under them. We hope that this book has adequately identified the dangers and opportunities involved in the untrammeled development of communications; we also believe that it demonstrates a very effective approach in American studies.

Grzegorz Kość and Krzysztof Majer

ACKNOWLEDGEMENTS

This collection of essays is based on presentations and discussions of the Annual International Conference of the Polish Association for American Studies, which was organized by the Department of American Literature and Culture, University of Łódź, and took place on November 7–9, 2007. The whole event would not have been possible without the support of the American Embassy in Warsaw, represented by Magdalena Wącisz, and that of Andrzej Dakowski, the Executive Director of the Polish-U.S. Fulbright Commission. Furthermore, we would like to thank the keynote speakers—David E. Nye of the University of Southern Denmark and Andrew Ross of New York University—as well as all the conference participants. Last but not least, we want to acknowledge our debt to the organizers, in particular Jadwiga Maszewska, Agnieszka Salska, Zbigniew Maszewski, Krzysztof Andrzejczak, Dorota Wiśniewska, Alicja Piechucka, Kacper Bartczak and Matthew Wilson.

Finally, special thanks to Beata Dąbrowa-Kość who helped with editorial work in the final stage of the project.

HOARDING THE GOLD: US TECHNOLOGY AND INTERNATIONAL TRADE

ANDREW ROSS

It has been said that technology is a civil religion in the US. When other belief systems are exhausted, or else confounded by realities on the ground that betray their ideals, we can always rely on the next generation of technology to deliver the dream of perfection to a fallen people, yearning for completion of things they have been promised. This kind of narrative is standard fare for American Studies, and it has enjoyed a straight line of continuity from Leo Marx's *The Machine in the Garden* (1964), which analyses the struggle between the Enlightenment legacies of pastoralism and technological progress, to David Noble's *The Religion of Technology* (1997), which records how successive generations of evangelists found redemption in the technological sublime. This is a field of literature that takes a nationalist frame for granted, and it is shamelessly exceptionalist, finding in American pragmatism a unique, homegrown response to the ideologies of progress associated with technical invention.

In recent years, however, this narrative has altered somewhat, and not just because scholarship in ASA has entered a post-nationalist phase. This narrative has fallen victim to the quickening forces of globalization, the influence and pace of which causes the ground to continually shift under our feet. Today, there is no easy congruence between the national interest and the technology-driven industries that were national champions, at home and abroad, during the heyday of the Cold War. The internationalization of production, which began in the 1960s with the offshore manufacture of radio and TV component parts, primarily in Japan and Taiwan, is now a properly global phenomenon—its subcontracting chains snake their way all over regions like East Asia, Eastern Europe, and Central America. Technology-driven corporations are hot to transfer overseas as many operations as they can as soon as possible. These are no longer limited to labor-intensive operations; they include R&D units which lie at the very top of the skills hierarchy, and which we have long been told would never leave American shores.[1]

The collapse of this last myth has been the hardest to swallow, since it goes to the heart of national pride about native innovation. Yet there are few obstacles that lie in the path of this kind of offshore knowledge and technology transfer—a temporary shortage of skilled labor in select regions of the world where such labor is conveniently cheap, or an immature legal regime for intellectual property in others. And then, of course, there is domestic public opinion, which is relatively easy to work up into a chauvinist lather, and indeed there has been much of that in recent years around the prospect of high-skill jobs going offshore.[2] Politicians take note of such sentiment, but Washington has little appetite for stemming the flow. For sure, there has been the spectacle of some senators from vulnerable states grandstanding on Capitol Hill, and on the campaign trail, and no elected official will resist the invitation, when called upon, to voice support for some sort of economic nationalism. But there has been no effective political will to overturn neoliberal trade policies which encourage the export of the nation's technological patrimony.[3]

Protectionist measures are generally only successful when nations are net importers—America's wall of high tariffs in the nineteenth century is one of the best examples of that during the era when the US was still the world's number one pirate nation. Today, even the national security hawks, so powerful in other aspects of foreign policy, are losing the battle to keep militarily sensitive technologies out of the hands of potential enemies.

For several postwar decades, there was a moratorium on exporting dual-use technologies to proscribed or suspect countries, whose militaries could absorb them to create mischief, or worse. With the ending of the Cold War, this moratorium was replaced with the more informal Wassenaar Arrangement, whose 33 member countries enjoyed considerably more individual discretion to control sensitive exports. This much looser agreement, combined with the loss of the U.S.'s commercial high-tech monopoly, meant that Washington could no longer effectively control technology denial to countries on its enemy list. In the meantime, the commercial trade in licensed exports shot up, and, by 2003, the export of general purpose microprocessors was entirely de-controlled.

---

[1] Walsh, *Foreign High-Tech R&D*; Saxenian, *New Argonauts*.

[2] Tonelson, *Race to the Bottom*.

[3] Bronfenbrenner, *Impact of U.S.-China Trade Relations*; Bronfenbrenner and Luce, *Changing Nature*.

Hardliners continued to insist that countries like China should be kept two generations behind in its access to advanced technology. On the other side, the powerful business wings of both political parties were acutely aware that China was already using its access to international technology and knowledge transfers to jumpstart its own high-tech industries, and that U.S. firms would lose vital opportunities if the more stringent controls were not relaxed. Most U.S. firms, and foreign investors in general, have come to accept that technology transfer, even the illegal appropriation of their own intellectual property, is part of the price of doing business in China. Indeed, the offer of technology transfer is a routine deal-maker for U.S. firms seeking better investment terms or market access. Most recently, the cost of access has been the establishment of a joint training or R&D center with a Chinese university or research institute. By dangling the promise of its vast domestic market, China has been able to attract the kind of technology and resources that it needs to upgrade its economy. By the end of 2004, Washington had bowed to the clamor from U.S. firms by softening its controls. An agreement on "end-use visits" was negotiated, and controls on the export of advanced semiconductor manufacturing equipment were eased.

What is the upshot of this relaxation of export controls? It is only a matter of years, perhaps months really, before we see a new generation of technologies whose American brand names will be their only effective relationship to the U.S. They will have been conceived, developed, tested, engineered, manufactured, and assembled in other parts of the world. So how does this fit with the vision of homegrown American know-how and innovation that has been one of the core elements of nationalist ideology for at least two centuries? Not very well. Will the result occasion any more significant public outcry, or government action, or will it be accepted as just one of the many contradictions that globalization has thrown our way? How much does it matter if no one in America knows any more how to make bicycles, let alone computers?

However you answer those questions, American Studies scholars should hardly be confounded by such developments. Historically speaking, literature in the field shows that the development of new industrial technologies has aroused more skepticism, anger and opposition among users than it has generated gratification among owners. The use of such technologies to deskill artisans, to speed up work tempos, and to enforce new divisions of labor designed to disempower operators is well understood and well-documented.[4] Overall, the argument that new

---

[4] Braverman, *Labor and Monopoly Capital.*

technologies promote efficiency in the workplace has generated much less evidence than the counter-claim that they are introduced to tighten control and surveillance over a workforce. This applies even in the age of Information Technology, where, among economists, the theory called the "productivity paradox" is still intact. No economist has yet come forward with statistical evidence that the introduction of IT into workplaces has boosted productivity appreciably (there are many in the economics profession who would be all too happy to deliver such evidence to the business leaders whose aspirations and belief-systems they do so much to embellish).

Adding to the longstanding anxieties about deskilling and surveillance have been even more sharply pronounced anxieties about technological disemployment—the fear that machines will displace our livelihoods entirely. A mainstay of the 1930s and 1960s, this fear was slightly revived in the 1990s during the dot.com boom. One of the buzzwords of that era was disintermediation. This term implied that digitization would cut out all the middle men who were necessary to distribute product in the brick-and-mortar world but who would be rendered redundant in the world of online business. Though it was overhyped, like everything else in the heady New Economy climate of that era, its consequences have nonetheless been felt in all areas of commerce as the practice of doing business online has spread. Ironically, one of the deepest impacts has been in the technology industries themselves, among employees who pioneered the new media technologies of that period.

One of the companies where I did ethnography for my book *No-Collar* had a rather famous slogan: "Everything that can be digital will be digital." The dotcommers who lived by that slogan were among those whose livelihoods suffered when the slogan morphed, over the next few years, into "Everything that can be outsourced will be outsourced." They had pioneered the technologies that made it technically possible to send their jobs offshore—web-based work-flow platforms that can break down jobs into discrete operations, allocate the work to different parts of the globe, and recombine the disparate portions into a meaningful whole that is serviceable to a project manager. They preached a business revolution in which they were the evangelical first movers but had little warning that the process would have such a huge impact on their own workplaces.

Even so, we do not hear much about technological disemployment these days. All of the relevant anxiety has been displaced into fears about job loss from outsourcing, and the blame, for the most part, is placed on foreign governments, (Beijing, for example) rather than on corporate CEOs who order the overseas transfers of jobs and technologies. Among these CEOs, there is a clear divide between larger companies that have been able to move offshore, and those who cannot, for reasons of their economy of scale, limited capital base, or the nature of the goods or services they provide. These smaller, domestically-focused producers have been losing business and profits to the China trade, primarily because they cannot compete with the so-called "China price" of imports. The National Association of Manufacturers was split down the middle on this issue for several years, with the small producers favoring self-interested alliances with the labor movement on China-bashing calls for protectionist measures, and the larger multinationals, who generally call the shots, favoring the free trade status quo. While this alliance generated a good deal of sound and fury on Capitol Hill, government agencies have followed the bidding of the multinationals, and are subject to the influence of the powerful China trade lobby that is controlled by the global firms.

In the US these days, we hear a lot of skepticism about the influence of the pro-Israel lobby, and deservedly so—cracks have begun to appear in the establishment consensus about unwavering support for Israeli policies.[5] The best evidence for this is that alternative views are now being so hysterically shouted down, even in the academy, where academic freedom has been trampled on, and appointments in Middle Eastern Studies roughly manhandled by the influence of pro-Israeli neo-conservatives. By contrast, not very much is heard about the China lobby. It is a different beast entirely, but arguably has a larger sphere of influence in trade policy than the Israel lobby does in foreign policy. The former exists for the purpose of increasing the scale and scope of US multinational operations in the PRC, and it is directed at home by the US-China Business Council, and by the American Chambers of Commerce in Shanghai and Beijing. The lobby's chief tool of political leverage is that the American business community in China is in a position to promote the liberalization of Chinese society and also to offer intelligence to diplomats to which they do not have access. On almost every aspect of trade and foreign policy in East Asia, the China lobby gets what it wants.[6]

---

[5] Mearsheimer and Walt, *Israel Lobby*.

[6] I watched this process in the course of researching my last book *Fast Boat to China: Corporate Flight and the Consequences of Free Trade; Lessons from Shanghai*. One of my primary research sites in Shanghai was the American Chamber of Commerce, where I spent a lot of time interviewing, schmoozing, and attending meetings. I was able to see how the offshore arm of the lobby rode out each wave of China-bashing to sweep over Capitol Hill, or, should I say, managed to turn it to advantage, because China-bashing is one of the most reliable ethnic instincts of white America.

What, then, is the role of technology in all this? Rather crucial. China is leaping up the technology curve much faster than Japan, Korea, or Taiwan did during their era of industrialization by invitation. Besides, China was not invited to industrialize by the advanced powers. It did so on its own dime and on its own terms, beginning in the Maoist era, when, contrary to the dominant iconography of Mao as a committed pastoralist, his focus on technology-driven industrialization produced many homegrown successes (some of them, especially in the area of high-tech, are not well-known at all). In the 1950s, the new communist state established a science and technology R&D network, modeled after the Soviet system, and its electronics arm went on to produce several generations of computers, in many cases with little or no gap behind the capitalist powers. China's first computer was developed in 1958, only one year after Japan's, and its first integrated circuit was produced in 1964, only five years behind the first U.S. patent. A microcomputer was developed by 1977 (even before IBM unveiled its PC), a microprocessor by 1980, and a supercomputer, along with an IBM-compatible PC, by 1983.[7] As in the U.S. and European cases, advanced technology was developed initially with the military end-user in mind, and a large cadre of sophisticated engineers staffed the thousands of research institutes and labs (many of them in the west of China, outside of the spotlight) which fed initiatives like the atomic and hydrogen bomb projects, along with a succession of satellite technologies. After the Soviet pullout in 1960, the goal of self-reliance in technological achievement became a matter of vast national pride.[8]

In the U.S., federal funding accounted for between 40 and 45% of high-tech R&D from 1958 through the early 1970s. By the late 1970s, the private sector's pace of development and funding commitment for commercial R&D had outstripped that of the defense systems.[9] By contrast with the US, China had virtually zero civilian or commercial demand, and so the government's research elites, often isolated from one another in their labs, had no incentive to pass on the innovations to would-be industrial exploiters. Efforts to stimulate civilian industry fell far short of expectations, especially since successful manufacturing in sectors like integrated circuits is so dependent on international expertise. So eventually, China's policy-makers looked to foreign talent and investment—mostly the Taiwanese—to establish a fully competitive industry in high technology.[10] Even so, they have done it on their own

---

[7] Lu, *China's Leap*, 6–9.

[8] Simon and Rehn, *Technological Innovation in China*, 1–19.

[9] Flamm, *Mismanaged Trade*, 36–37.

[10] Hsing, *Making Capitalism in China*.

terms. The result today is a burgeoning semiconductor sector that is not very far behind the industry leaders, and the growth of a cheap skilled labor pool that makes the PRC all the more attractive to multinationals.

China is unique in this regard because no modern industrializing country has been able to compete for top-end jobs at the same time as it effortlessly absorbs jobs much lower down in the production chain. To command this spread—from the lowest assembly platform work to the upper reaches of industry and services—is to be in a position to set the global norm for employee standards as no country has before. Given the chronic disregard for job security and workplace rights in China's foreign-invested private sector, such a norm is a clear threat to the stability of livelihoods everywhere. I refuse to accept that this is a threat hatched in Beijing. If China did not provide the most currently profitable mix of authoritarian governance, cheap, abundant labor, and investor-friendly policies, it would be sought out elsewhere. Though such conditions would not exist without government cooperation, the primary instigators and beneficiaries are global corporations. They stand to profit most from the normalization of an environment where jobs, capital, and technology can be transferred at a moment's notice and with complete impunity. If corporations cannot be held to consistent standards, responsibilities, or respect—and the working reality of offshore free trade does not demand any of these—then ordinary people in every corner of the globe will be at a loss to salvage any control over their futures and that of their communities.

Technology transfer is crucial to the US-China trade, and the mutual dependence of the US and the Chinese economies makes this transfer inevitable. The only real counterweight to this process from a US perspective lies in the realm of military technology, which Washington has made available in large quantities to Taiwan over the years. The more recent penetration of US military into Central Asia has resulted in the virtual encirclement of China, and Washington's cozying up to India has not helped to allay Beijing's fear of being encircled. When Secretary of State Condoleezza Rice declared, in a visit to Delhi in March 2005, that "the US policy was to help India become a major world power in the 21st century," the momentous implication was that one of the two emerging Asian mega-economies would be used to counterbalance the influence of the other. No major power in history has ever declared their intention to create another superpower. The subsequent India-US nuclear co-operation agreement sent a clear signal about how far Washington was prepared to go down the path of technology transfers to India in an effort to contain China and retain its ability to control the balance of power in the region.

The other weapon that Washington holds in its arsenal is not so much a technology per se as the legal regime which governs the licensed use of technology. I am referring to intellectual property, which is the number one tool available to US trade negotiators when they sally forth to promote the interests of US multinationals. In a globalized economy, there is little chance of stopping technology transfer, but a legal regime that regulates its licensed use, and which thereby guarantees a durable revenue stream, is the next best thing. This is why the drive to harmonize IP laws of all WTO members is so important to Washington. But does it correspond, in any way, to protecting the national interest? The case is even more dubious than with the technologies themselves. These days, many American universities hold patents and copyrights, and so one could argue that the revenue somehow benefits students. Nevertheless, this benefit has to be balanced against the social cost of commercializing the results of academic research, which are considerable, and which has generated a sharp debate in its own right. Most of the commercially valuable IP lies in the hands of large corporations, and in the US the stampede to secure more of it has seen the biggest property grab since the Treaty of Guadalupe.

The more common analogy used by campaigners against the IP monopolists is with the rapacious land enclosures of the 16th, 17th and 18th centuries in England, when land used as a commons was deeded exclusively to private owners. Resistance to the enclosure of the information commons is a rallying cry of the loose coalition of groups and organizations that have sprung up to contest the rush to propertize virtually everything under the sun.[11] (My favorite example is that Donald Trump is trying to trademark the expression "You're Fired" used on The Apprentice TV show, along with the hand gesture that accompanies it.) This coalition is packed with lawyers eager to take on monopolists in the courts, and artists unwilling to see their creative professions cede to full marketization. It includes high-skill engineers in the technology industries, whose professional labor ethos is opposed to any proprietary hold on information technologies, and who constitutionally lean toward the hacker ethic of shareware. And it includes overseas activists, many with advanced science degrees, fighting the efforts of multinationals like Syngenta, AstraZeneca, DuPont, Monsanto, Merck, and Dow to patent seeds, livestock, plant genes, and other biological raw materials that have been the basis of subsistence farming in the developing world for centuries.

---

[11] Lessig, *Future of Ideas*; Lessig, *Free Culture*; Litman, *Digital Copyright*; Perelman, *Steal This Idea*; Boyle, *Shamans, Software, and Spleens*; Bettig, *Copyrighting Culture*; Vaidhyanathan, *Copyrights and Copywrongs*; and von Lewinski, *Indigenous Heritage*.

What is distinctive about the disparate members of this loose coalition is the high degree of brainpower involved. They are drawn from sectors that are key to capital accumulation in the new knowledge economy, and collectively, they represent a significant, dissenting fraction of the cognitariat.

In this respect, they constitute what Marx himself referred to as "the general intellect," in an obscure "Fragment on Machines" in the *Grundrisse*, a passage which has attracted a good deal of attention from the post-*operaiste* Italian school of autonomists.[12] Marx, of course, was as ambivalent about technology as any of us, but in this passage, he predicts that capitalism's dependence on science would increase and that the generation of profit would depend less on direct labor time and more on the harnessing of mental powers and knowledge resources—"the general productive forces of the social brain."[13] Technology, in the form of fixed capital, would be the most efficient way for owners to coordinate and absorb mental labor. Yet to the degree to which the general intellect is a collective entity, production would become more and more social in nature. Ever alert to evidence of the bourgeoisie digging its own grave, Marx imagined that this latter development might lead to the dissolution of wage labor and private ownership along with capital itself.

That moment is not yet upon us, but it is plausible to conclude that the conflict over IP is, in large part, a consequence of the potential harbored within the general intellect. Efforts to administer an effective division of labor within the knowledge industries increasingly depend upon control over the IP inside employees' heads and the capacity to affect "knowledge transfers" without too much friction. Information monopolists have undertaken a massive property grab to prevent leaks in the system. But the leaks are being sprung nonetheless, and the Internet, which teems with unauthorized content, is the most porous of all public entities. Corporate managers, bent on disciplining rogue users through the use of electronic locks or other forms of Digital Rights Management (DRM) are now in a running battle with the ever-proficient hackers of the technocratic fraternity. Punitive policing of illegal downloaders among the general population runs the risk of adding to the record of case law that supports fair use and casts doubt on the legitimacy of all-out privatization. And, overseas, rent-seeking multinationals are running into significant resistance in their efforts to drive home global IPR harmonization.

---

[12] Virno and Hardt, *Radical Thought in Italy*; Dyer-Witheford, *Cyber-Marx*.

[13] Marx, *Grundrisse*.

No one can doubt that these coercive efforts will continue apace. IP-driven industries—from microchips and biomedicine to multimedia entertainment—stand at the commanding heights of a rapidly globalizing economy, and their owners are bent on hammering out a property regime that will keep them there for decades to come. As always, their ability to shape and create their own opposition makes it easier to recruit their enemies. At least two generations of hackers have agonized over accepting lucrative offers of employment within corporate or government IP security. To cite another example, the Open Source movement in software is now integral to the business of multinationals, both as a source of free labor and a qualitatively superior source of technical competence.

These skirmishes are the result of intra-class conflict. The key to a more sustained challenge to knowledge capitalism lies in cross-class coalitions. For example, the march of insecurity into the upper reaches of industry and services has meant that skilled employees increasingly share the experience of contingency, if not disposability, with unskilled, low-wage or blue collar workers who felt the first brunt of corporate flight some time ago. In this development, some theoretical commentators from the Italian school see the formation of a multi-class *precariat*, linked by joint concerns and capable of developing a unity of consciousness and action on an international scale. Critics of this view dismiss as naive the assumption that highly trained aristocracy of labor will find common cause with the less skilled, simply on the basis of insecurity.

Yet we cannot afford to reject out of hand any evidence of, or potential for, these forms of identification. In my own research, for example, in IT and other technology-driven firms, I have found it common for employees to refer to their workplaces as "high-tech sweatshops," especially when they are pressured by long hours, deadline speed-ups, and divisions of labor that reduce employee autonomy. No doubt, these are throwaway comments, and are often simply expressions of the most cynical side of office humor, but they also contain real elements of self-recognition and identification with the plight of those toiling in workplaces customarily associated with sweatshop labor. "Wage slavery" once resonated as a slogan, in the 1840s, for skilled artisans facing down factory deskilling, and it also played a role in Abolitionist sentiment and action.[14] Today's anti-sweatshop movement has helped to revive public sympathy for the predicament of workers in labor-intensive jobs, but it has also provided a moral language and posture for those in value-added trades who see their

---

[14] Roediger, *Wages of Whiteness*.

own occupational sectors following a similar path, offshore and downmarket.

One of the more conspicuous examples is the case of the academic professional. Once a domain of occupational security, higher education is now awash with contingency; between a half and three quarters of its teaching has been casualized, leaving a minority in the tenure-stream to exercise the security and the academic freedoms that are the signature of a free society. Yet the experience of deprofessionalization has triggered an embryonic labor movement that may yet transform the workplace if it can successfully draw in larger numbers of the securely tenured.[15] The concomitant demystification of academe and its genteel cult of disinterestedness has cleared the way for a more accurate assessment of its worklife—an advance in consciousness that will bear fruit in action.

In addition, the erosion of job security and work standards have forged a new perspective on the conditions of those who never enjoyed such advantages in the first place, whether because of their gender, racial background, or regional location. The breadwinner of the Cold War "family wage" breathed a different air from those employed in the secondary labor markets of the era, and often did so at the cost of the latter. Today, the casualized features of the latter are more familiar to all of us, whether in high-wage or discounted occupational sectors. Such circumstances make it more difficult to ignore the principle that justice for one must be pursued as justice for all.

Works Cited

Bettig, Ronald V. *Copyrighting Culture: The Political Economy of Intellectual Property*. Boulder, CO: Westview Press, 1996.

Bousquet, Marc. *How the University Works: Higher Education and the Low-wage Nation*. New York: New York University Press, 2007.

Boyle, James. *Shamans, Software, and Spleens: Law and the Construction of the Information Society*. Cambridge, MA: Harvard University Press, 1996.

Braverman, Harry. *Labor and Monopoly Capital: The Degradation of Work in the Twentieth Century*. New York: Monthly Review Press, 1998.

[15] Bousquet, *How the University Works*; Krause et al., *University Against Itself*; Martin, *Chalk Lines*; Nelson, *Manifesto of a Tenured Radical*; Johnson, Kavanagh, and Mattson, *Steal This University*.

Bronfenbrenner, Kate L. *Impact of U.S.-China Trade Relations on Workers, Wages, and Employment*. Washington, DC: U.S.-China Economic and Security Review Commission, 2001.

Bronfenbrenner, Kate L., and Stephanie Luce. *The Changing Nature of Corporate Global Restructuring: The Impact of Production Shifts on Jobs in the U.S., China, and Around the Globe*. Washington, DC: U.S.-China Economic and Security Review Commission, 2004.

Dyer-Witheford, Nick. *Cyber-Marx: Cycles and Circuits of Struggle in High-technology Capitalism*. Urbana: University of Illinois Press, 1999.

Flamm, Kenneth. *Mismanaged Trade? Strategic Policy and the Semiconductor Industry*. Washington, DC: Brookings Institution Press, 1996.

Hsing, You-tien. *Making Capitalism in China: The Taiwan Connection*. Oxford and New York: Oxford University Press, 1998.

Johnson, Benjamin, Patrick Kavanagh, and Kevin Mattson, eds. *Steal This University: The Rise of the Corporate University and the Academic Labor Movement*. New York: Routledge, 2003.

Krause, Monika, Mary Nolan, Michael Palm, and Andrew Ross, eds. *The University Against Itself: The NYU Strike and the Future of the Academic Workplace*. Philadelphia: Temple University Press, 2008.

Lessig, Lawrence. *Free Culture: How Big Media Uses Technology and the Law to Lock Down Culture and Control Creativity*. New York: Penguin, 2004.

—. *The Future of Ideas: The Fate of the Commons in a Connected World*. New York: Random House, 2001.

Litman, Jessica. *Digital Copyright Protecting Intellectual Property on the Internet*. New York: Prometheus Books, 2001.

Lu, Qiwen. *China's Leap into the Information Age: Innovation and Organization in the Computer Industry*. New York: Oxford University Press, 2000.

Martin, Randy, ed. *Chalk Lines: The Politics of Work in the Managed University*. Durham, NC: Duke University Press, 1999.

Marx, Karl. *Grundrisse: Foundations of the Critique of Political Economy*. Translated by Martin Nicolaus. Harmondsworth: Penguin Books, 1973. http://www.marxists.org/archive/marx/works/1857/grundrisse/ch13.htm#p694.

Mearsheimer, John J., and Stephen Walt. *The Israel Lobby and U.S. Foreign Policy*. New York: Farrar, Straus, and Giroux, 2007.

Nelson, Cary. *Manifesto of a Tenured Radical*. New York and London: New York University Press, 1997.

Perelman, Michael. *Steal This Idea: Intellectual Property Rights and the Corporate Confiscation of Creativity*. New York: Palgrave Macmillan, 2002.

Roediger, David. *The Wages of Whiteness: Race and the Making of the American Working Class*. New York: Routledge, Chapman, and Hall, 1999.

Ross, Andrew. *No-Collar: The Humane Workplace and Its Hidden Costs*. New York: Basic Books, 2002.

—. *Fast Boat to China: Corporate Flight and the Consequences of Free Trade; Lessons from Shanghai*. New York: Pantheon, 2006.

Saxenian, Annalee. *The New Argonauts: Regional Advantage in a Global Economy* Cambridge, MA: Harvard University Press, 2006.

Simon, Denis Fred, and Detlef Rehn. *Technological Innovation in China: The Case of the Shanghai Semiconductor Industry*. Cambridge, MA: Ballinger Publishing Co., 1988.

Tonelson, Alan. *The Race to the Bottom: Why a Worldwide Worker Surplus and Uncontrolled Free Trade Are Sinking American Living Standards*. Boulder, CO: Westview Press, 2000.

Vaidhyanathan, Siva. *Copyrights and Copywrongs: The Rise of Intellectual Property and How It Threatens Creativity*. New York: New York University Press, 2002.

Virno, Paolo, and Michael Hardt. *Radical Thought in Italy: A Potential Politics*. Minneapolis: University of Minnesota Press, 1996.

von Lewinski, Silke. *Indigenous Heritage and Intellectual Property: Genetic Resources, Traditional Knowledge and Folklore*. The Hague: Kluwer Law International, 2004.

Walsh, Kathleen. *Foreign High-tech R&D in China: Risks, Rewards, and Implications for US-China Relations*. Washington, DC: Henry L. Stimson Center, 2003.

THE GOLDEN AGE OF AMERICAN RADIO: ENTERTAINMENT, NEWS, AND POLITICS IN THE UNITED STATES, 1926 UNTIL 1956

DAVID A. JONES AND JOANNA WALUK

Our paper will focus on the emergence and development of Radio as a Big Business and as a form of both commercial and political communication in the United States. We shall advance our theory of how and why the radio came to be so significant in "nation building" of the United States during its Great Depression and the years leading up to it, during, and after World War Two. We have coined the word and concept which we call "Memorideas" to link radio content to American Nationalism through the medium of Radio during its thirty-year period that has become known as Radio's "Golden Age."

Memorideas and American Nationalism

The United States of America is a "melting pot" population who shares no common ethnicity. Although heavily Germanic by heritage in many geographical areas of America (especially the Midwest), with concentrations of Irish and Italians in major urban areas, Americans in reality come from almost everywhere. For this reason, primarily, it is difficult to detect a natural nationalism within America. Some historiographers have argued that America evidences no nationalism whatsoever, for better or worse. This paper argues that they are wrong. American Nationalism does exist, but it had to be constructed, and was, by the radio during its "Golden Age." Although the radio as a machine was indispensable to this end, machinery was not what constructed American Nationalism. That was done by the character of radio programming and by the character of those who did that programming.

The key to American Nationalism in the Depression of the 1930s was what we call "Memorideas." Neither memories nor ideas, they are essentially a combination of the two. Memories are individual. "Memorideas" instill memories differently from one person to the next, but more or less identically across the listening population of Radio. Nationalism requires a commonality: in experience, in emotion, in values, and in a sense of what is important. What Radio [initial capital indicates Radio as an intellectual medium] did was to create and disburse nationally, then globally, a set of local memories that were transformed into giant ideas. First, Radio had to reach the nation. Next, it needed to reach other nations. In reaching the world, American Radio came not only to spread American "Memorideas" to foreign populations, but to Americans living outside of America, particularly those posted overseas in the American armed services, as World War Two approached. "The Cavalcade of America" typified this genre of radio programming. The hallmark of "The Cavalcade" was hero worship: a plethora of heroes as different from one another as night and day. They ranged from the Marquis de LaFayette to Abraham Lincoln, from Annie Oakley to William F. ("Buffalo Bill") Cody. Native Americans such as Chief Geronimo and Chief Sitting Bull were included alongside Supreme Court Justice Oliver Wendell Holmes, Jr.[1] Some have called this complex assortment of Radio programs "Total Theater," in reference to Richard Wagner's concept of a *Gesamtkuntswerk*, or "Total Art Work," in which music, drama, and dance operate together.[2]

Broadcasting as "Big Business"

The inventor of the radio, Chevalier (Knight) Guglielmo Marconi, formed American Marconi Corporation in 1897. Under this umbrella, "commercial" radio technology was perfected and advanced. Upon purchasing United Wireless Company in 1912, American Marconi dominated commercial radio broadcasting in his new homeland. When the United States entered World War I, all amateur ("Ham") radio operations were shut down. Most commercial radio closed as well. Some U.S. government officials, mainly connected with the Navy, planned to create a government monopoly over American commercial radio by purchasing Federal Telegraph Company and a majority of the radio stations affiliated with American Marconi. When Congress learned of this after the war, the Navy was forced to return radio broadcasting assets to their original owners.

After the war, also, radio broadcasting became attractive to three companies that were interested in manufacturing radios: American Telephone

---

[1] See "The Cavalcade of America: Myth and Reality, Hero Worship in American Radio" in American Studies at the University of Virginia Program Guide.

[2] Packer, Randall, and Jordan, "Pioneers."

and Telegraph Company ("AT&T"), General Electric ("GE"), and Westinghouse. The first local radio station in America was KDKA in Pittsburgh, home of Westinghouse. Within a year, Westinghouse opened three more local stations: WBZ in Springfield, Massachusetts, KYW in Chicago, and WJZ in Newark, New Jersey. In 1919, the U.S. government pressured American Marconi to sell its assets to GE. Shortly thereafter, GE formed Radio Corporation of America ("RCA") which became the pioneer in national broadcasting. RCA made its first national broadcast on 2 July 1921, when it covered the boxing competition for the title of World Heavyweight Champion that Jack Dempsey won over Georges Carpenter. From then on, the American public listened to sports on the radio, particularly on weekends, including baseball, America's national sport. Before World War II, major college as well as professional football teams plus professional hockey and some basketball games received local or national radio coverage.

In November, 1921, KYW in Chicago broadcast complete operas. In February, 1922, WJZ in Newark aired entire Broadway musicals performed by the Broadway cast. In August, 1922, WGY in Schenectady, New York began to broadcast full length stage plays. Clearly, the interface of drama and music had begun on radio, in addition to sports, news and weather. Regulation seemed needed.

On 1 December 1921, the U.S. Bureau of Navigation adopted regulations that allowed entertainment to be broadcast on wavelengths of 360 meters (833 kilohertz), and market and weather reports to be broadcast on wavelengths of 485 meters (619 kilohertz). The year 1922 witnessed great leaps forward for radio. During that year, AT&T opened a local radio station in New York City, WEAF (presently WFAN), and leased out its content to producers who paid to broadcast. This became known as "toll broadcasting" and was the antecedent of today's paid sponsorships, known as "commercials." By 1922, daily broadcasts across much of America included news and music, with sports on weekends.

In 1926, radio manufacturers GE, RCA, and Westinghouse bought out AT&T networks and formed the National Broadcasting Company ("NBC"). Eventually RCA bought out the others under the leadership of its longtime chairman, Brig. Gen. David Sarnoff. Then began what Arthur Anderson has called "The Golden Age of Radio." Sarnoff, himself a Jewish immigrant from Belarus, gained national fame as a broadcaster much later when he personally covered Nazi Germany's invasion of Poland on 1 September 1939.

In September 1928, William S. Paley bought Columbia Broadcasting System ("CBS") with 16 failing stations and turned it into what is called "the Tiffany network." Paley did this during the 1940s by "raiding" NBC and offering its radio performers much larger salaries and stock options than the competition. The news star of CBS was commentator Edward R. Murrow. Eventually, CBS had contracts with 28 of America's leading sports franchises including the National Football League (NFL), National Basketball Association (NBA), National Hockey League (NHL), and key baseball teams, including New York's Yankees and the Chicago White Sox. By 1951, CBS aired 14 of America's top nighttime entertainment listings including Jack Benny, Gene Autry, Gary Moore, plus George Burns and Gracie Allen. Also, CBS aired Arthur Godfrey on its morning program.

During the Great Depression, also, Roman Catholic "radio priests" emerged with a combination of religious and political conversation. At the time, the most famous was the Rev'd. Charles Coughlin, who was blamed for anti-Semitic remarks. Over time, the Most Rev'd. Fulton J. Sheen, Archbishop of Rochester, New York, grew to a stature larger than Father Coughlin's, with over four million weekly listeners by 1950.

President Franklin Delano Roosevelt utilized radio as a forum for announcing to the American public the goals and progress made by his Administration in the field of economic recovery. This he did with "Fireside Chats" that began on 12 March 1933, only nine days after his Inauguration. Arguably, FDR's most important economic Fireside Chat was entitled "On the Purposes and Foundations of the Recovery Program" and delivered on 24 July 1933. President Roosevelt broadcast 30 Fireside Chats over nearly 12 years from 1933 until 1944.[3] These discussions were broadcast on different days (at least one on every day of the week) but almost half (12) on Sundays. They lasted from 12 to 40 minutes. Through these broadcasts, FDR seemed to bond tightly with the American people, as if they were talking face to face.[4] Perhaps this is one reason why Americans received President Roosevelt's Message to Congress on 8 December 1941, the day after Japan's attack on Pearl Harbor, Hawaii, with such emotion and commitment to the Allied cause. To a lesser extent, perhaps, but still with patriotic emotion, the American public listened to President Harry S. Truman on radio as he explained his reasons for the atomic bomb attacks on Japan at Hiroshima and Nagasaki, and then as he announced and described the Japanese surrender aboard the U.S.S. Missouri in Tokyo Bay.

---

[3] Mid-Hudson Regional Information Center, *Fireside Chats*.

[4] Ibid.

Between 1938 and 1941, the Federal Communications Commission (FCC) began to investigate NBC and CBS as monopolies. It chose to focus on NBC because the company operated two networks, called NBC "Blue" and NBC "Red." Facing a divestiture order, NBC agreed to sell its control of NBC "Blue" to American Broadcasting System ("ABC"), owned by lifesaver tycoon Edward J. Noble. ABC never occupied a large market share of American radio during the "Golden Age of Radio." In fact, it faced bankruptcy until 1954 when it signed a contract with Roy and Walt Disney to broadcast "Disneyland" cartoon characters on its network.

The focus of this paper is on the content of American radio programs before television came of age. As such, it may be subdivided into four program sections: (1) comedy, drama and music, quiz shows; (2) sports; (3) news and weather; and (4) politics. In addition and necessarily, this paper will focus on the sponsors of commercial radio programs, together with the content, drama, and music of their commercial advertisements ("commercials"). All of these changed across commercial radio's "Golden Age" from 1926 until 1956 when commercial television emerged to replace the radio as the medium of America's daily entertainment.

Fascist Japan attempted to use American radio for its own propaganda purposes with various female broadcasters. Frankly, Americans considered this effort to be comical, and the end result was to inspire in most Americans the sense that Japan was both childish and evil, an adversary that American Fitness, Innovation, and Technological superiority could easily overcome. Although America defeated Japan, the road to victory was not easy. One of America's responses to "Tokyo Rose" was George Kaner's "Tokyo Moshe" who ridiculed the "Tokyo Rose" broadcasts with his theme song, set to the tune of "London Bridge Is Falling Down." "Tokyo Moshe" became hugely popular after the war among the youth of Japan.

At the war's end, a Japanese-American woman named Iva Toguri D'Aquino was framed for treason by the FBI and accused of being "Tokyo Rose." In fact, no such person existed, and American military personnel created that label to refer to the female Japanese broadcasters collectively. Radio is partially responsible for her false imprisonment and spurious conviction. Worse, radio was largely responsible for the American refusal to allow her to return to the United States for the birth of her child, who consequently died. American radio commentator Walter Winchell used his influence against D'Aquino, who served six years in prison but was eventually pardoned by President Gerald R. Ford.

On 24 February, 1942, the United States government began to broadcast the Voice of America ("VOA") across Europe and China. To this Radio Free Europe ("RFE") was added in 1950. Both of these networks offered feature programs, news, and political commentary. VOA began each daily broadcast with the music: "Columbia, the Gem of the Ocean," and played the American Revolutionary War song, "Yankee Doodle Dandy," at regular intervals. RFE and VOA were designed to provide international information to promote democratic values in countries across the world.[5]

The Content of Radio Programming

Radio programs during the "Golden Age" tended to focus on four areas: (1) comedy, drama and music, quiz shows; (2) sports; (3) news and weather; and (4) politics. Each deserves separate discussion, because the genres and the broadcasters were almost entirely different. Yet, these different genres converged in the form of what we term "Memorideas" to forge a strong American Nationalism.

Comedy, Drama, Music, Quiz Shows

One must not forget that Radio entered what we now call its "Golden Age" around 1926, and that this was only some three years before the beginning of the Great Depression. The mood of America was downcast, approaching *anomie* and despair. In fact, social upheaval and a battle between or among social classes were quite possible, as was the case in 1917 Russia or in China during its "Cultural Revolution" between 1966 and 1977. Chaos and public disorder were significant possibilities at least from 1929 until 1937. One cannot say for certain that all this was prevented solely by the Radio. What one can argue is that Radio brought a new type of entertainment that gave the American people hope. The Nationalism that emerged would serve America and Americans well during World War Two.

This occurred in at least two ways. Firstly, the Radio brought access to culture across the barriers of class. The rich and the poor alike could suddenly listen for free (or, more accurately, for the price of a radio that could be a shared investment of extended family units) to musical and theatrical performances theretofore enjoyed primarily by the upper classes and the bourgeoisie, largely in major cities such as Boston, New York, and Philadelphia. Secondly, Radio bridged the gap between city and countryside, functionally leveling American society by geography within

---

[5] Short, *Western Broadcasting.*

class. No longer would country folk have to travel to a major city to hear relatively high caliber performances, both with and without music.

By 21st-century standards, to be sure, Radio broadcasts were funky: the sound was somewhat distorted and had the ring of a tin can, and, depending upon technical factors (such as broadcast quality, radio quality, air wave frequency, and type of surrounding terrain as well as the weather) clarity of reception could be interrupted by "noise." Also, theatrical shows were quaint by our current standards, the jokes rather primitive and sometimes insulting, the dialogue sometimes racist or hurtful in other ways, and of course tied directly to the issues of the day. One of many examples would be the long-running "Amos 'n' Andy Show" with two white men pretending to be black.

Several other teams of comedians reached national (eventually global) fame with their serial radio shows: George Burns and Gracie Allen ("The Burns and Allen Show"), Lucille Ball and Desi Arnaz ("I Love Lucy"), Ozzie and Harriet Nelson ("The Ozzie and Harriet Show"), all married couples at the time.[6] Some comedians blended comedy with music, such as Jack Benny who played his trademark violin. These and other radio comedians provided respite from real life hardship and tragedy at intervals that diminished from a week to an evening. They also portrayed women in stereotypical roles that may not seem appealing today. Gracie Allen "played dumb," Desi Arnaz seemed to bend over backwards as a Cuban-American husband to please his red-haired wife. Ozzie and Harriet seemed too good to be true, with Harriet devoting all her efforts to housekeeping for their two sons, raising them in a typical bourgeois neighborhood of America's 1950s.

Afternoons were dominated by situation comedies ("Sitcoms") known as "soap operas," named to describe their sponsors, almost invariably companies such as Colgate, Palmolive or Proctor and Gamble, which produced soap, laundry detergent, toothpaste, and dishwashing detergent. Clearly, "soap operas" and the companies that sponsored them were aimed at attracting an audience of adult middle class women who did not work but stayed at home, the trend of the times.

Drama tended to be surreal, in part at least, whether it be Detective, Mystery, or the very popular Western genres. That Americans may have been too gullible in believing much of what they heard on Radio is confirmed by the near panic that resulted from the 1938 airing of Orson Welles's version of H. G. Wells's *War of the Worlds*. Thousands of

[6] Lucille Ball and Desi Arnaz eventually divorced, but after Radio's "Golden Age."

cosmopolitan New Yorkers really thought that "Martians" had invaded the New Jersey coast, and fled to inland shelters in mass panic. This evidences the "hypnotism" that Radio exerted on its listening audience during its "Golden Age."

Radio music was far less distorted than drama, but even contemporary critics sometimes called it "sentimental"; much of it, with the exception of professional opera, would not be successful today. The compositions were at least as important as the stars who sang them. Indeed, probably, music made the stars famous, rather than the reverse. Such was the case with at least two songs by Irving Berlin, himself a newcomer to America: "God Bless America," often sung by Kate Smith, and the song that sold more records than any other ever, "White Christmas," made famous by Bing Crosby. Berlin was Jewish, and his music broadcast across America by radio demonstrated the interblending of religious faiths in secular America during the late prewar period, itself a bedrock of the emerging American Nationalism.

One genre that is still timely is American "Country" music that really began to go national with the radio, particularly from Nashville, Tennessee's music hall known as the "Grand Ole Opry." Texas stars such as Bob Wills and Hank Williams, Sr. began their Nashville careers along with Roy Acuff, Eddie Arnold, and "Tex" Ritter. In some ways, Roy Acuff's "Wabash Cannonball" (about a train locomotive) became a national legend on radio. Along the way, it was joined by Woody Guthrie's "This Land Is Your Land" (February 1940). Together, these songs crafted a powerful respect for the United States as a Nation among the American people, many of them relative newcomers, largely from Europe. Guthrie was a "hobo" or man of the road much of his life. He saw what few Americans had seen with their own eyes, that is, American diversity, and he reported it on radio in his songs. One could argue that Woody Guthrie's songs, together with Radio as their medium, contributed significantly to the crafting of fighting teams during World War Two and to the eventual racial desegregation of the American armed services.

Popular songs and superstar singers made their national debuts on American Radio. These included "Big Bands" such as those led by Glenn Miller and Tommy Dorsey, together with the singers whom they made famous, such as Frank Sinatra and Louis Armstrong.[7] Bing Crosby and Perry Como made frequent radio appearances, as did their records. On 9 September 1956, Elvis Presley sang his "Love Me Tender" for the first

[7] Louis Armstrong, an African American, was also an outstanding trumpet player for the Big Bands.

time on CBS Radio's "The Ed Sullivan Show," shooting at that very moment into international stardom as "The King" of Rock 'n' Roll music. That show was broadcast at the same time live on CBS television, but most Americans did not yet have a TV set. Inside of a year or two, more or less, they would, and radio would be relegated to the car. From about the same period, average-priced automobiles began to have a workable radio.

Radio broadcasts contained their fair share of quiz shows. These ranged from country "spelling bees," usually with children as contenders, through questions concerning sports trivia to the historical and the political. The show "Information Please" tended to include celebrity quiz contestant Oscar Levant, an almost insane genius, brilliant pianist, friend of George Gershwin, and probably the best performer of Gershwin's "Rhapsody in Blue" on the grand piano in history. Later Groucho Marx hosted the quiz show "You Bet Your Life" from 1947 through most of the 1950s.

The most successful radio programs tended to be known as "variety shows." America's all-time biggest recording star, Al Jolson, finally came to host the "Kraft Music Hall" from 1947 on into the 1950s. From the 1950s, America's most popular variety show—first on Radio, then on television—was CBS' "The Ed Sullivan Show," hosted by Sullivan, who built his reputation first as "the nationally syndicated columnist" for the *New York Daily News*. Sullivan hosted the debuts of Elvis Pressley and the Beatles, among many other stars. Variety shows brought stars and emerging stars to the forefront of public attention. Many performers were members of minority groups, ethnically or racially, and some were female. By listening to Radio, the American people gained respect for the talents of individuals, set apart from their physical appearances, ethnicity reflected in surnames, religious faith, and gender. Radio was an equalizer.

Sports

Sports did what comedy, drama, music and quiz shows really failed to do completely: united Americans across racial barriers. From Joe Louis on, most of the greatest stars of boxing, for example, were African American. Several contemporary commentators declared that the media demanded the entry of Blacks into boxing, sensing the lucrative stakes for promoters and the media, resulting from more powerful fighters and black audiences.

Once Jackie Robinson broke baseball's color barrier, the segregation of sports players was over and numerous black baseball players joined major league teams. For American men, especially, weekend afternoons on the radio were dominated by sports, particularly by baseball during the "Golden Age." Baseball teams were organized within one of two "leagues": the American League and the National League. Perhaps as much as, or more than, any other annual event during Radio's "Golden Age," American Baseball's World Series brought together Americans representing every demographic variable. The key to understanding this concept is the fact of a "tomorrow": that next year or the year after that, another team (perhaps nearer to home) would win the league's "pennant" and go on to try to win the World Series championship. In fact, America's largest city hosted its most successful baseball team: the New York Yankees. The success of the Yankees in winning the American World Series championship many times diverted popular attention away from the horrors of poverty, looming warfare, "Jim Crow" laws, crime and rivalry between political candidates and parties.

Some college football games were broadcast on radio, also, especially in big college towns such as Notre Dame's South Bend, Indiana. Horse racing and harness racing were very popular radio sports, particularly the "Triple Crown" races in spring: The Kentucky Derby, Preakness, and Belmont Stakes. Another notable example were the Olympic finals, beginning with African American runner Jesse Owens's immortal Gold Medal victories in Berlin in 1936 which flew in the face of Hitler's racism.

News and Weather

The radio airwaves were shared by the three giant national networks, together with a myriad of local radio stations. News and weather, normally, were local programs, often aired in between the other forms of entertainment broadcast by the networks. They came on the hour, and, in less detail, also on the half hour. Their sponsors were different as well, frequently being local stores or car dealers instead of national manufacturers.

Politics

As President of the United States, Herbert Hoover spoke on the radio at least once. It was his successor, Franklin Delano Roosevelt, who began to optimize the use of radio as a tool with which to communicate directly with the American people. His inaugural speeches were broadcast on the radio, but these were surpassed by his "Fireside Chats," during which he brought important issues directly to the American people. These he delivered every several months over the course of his 12 years in office.

Frankly, he mesmerized his constituency almost as if they had enjoyed an opportunity to meet him face to face or shake his hand.

In fact, radio helped immensely to disguise FDR's "splendid secret," as he called it, the fact that most of the day he was wheelchair-bound, paralyzed on account of polio. Almost none of the American public knew anything about Roosevelt's handicap, a fact that would have surely been impossible to hide from television cameras a generation later. Cinema photography portrayed Roosevelt at work sitting in an office chair behind his desk, driving his "Famous Ford" convertible car that had hand controls invisible to cameras, or walking amidst a throng of politicians in Congress, his crutches or braces invisible behind several bodyguards who appeared simply to be holding his arms, shielding him from crowds. On the radio, all this was distant. FDR's voice was intimate. More and more Americans came to trust FDR amidst his radio campaign song: "Happy Days Are Here Again."

Thus, Americans prepared for war rather enthusiastically when their President, appearing live as he did for his Fireside Chats, spoke before a joint session of Congress on 8 December 1941, the day after the Japanese attack on Pearl Harbor, Hawaii. In this address Roosevelt masterfully asked for a retroactive declaration of war, using his heavy Massachusetts accent, apparently acquired during his years at Groton School, followed by Harvard. As FDR's wartime ambassador to Great Britain, John Gilbert Winant, wrote when he sent FDR George Washington's walking stick as a present in 1945, "more than anything else, Mr. President, you symbolize *courage*." Such was the Roosevelt image to almost all Americans alive during the 1940s, transmitted on radio in a voice that did not wither with age or falter with stress.

The Content of Radio Commercials

In addition and necessarily, this paper also has to focus on the sponsors of commercial radio programs, together with the content, drama, and music of their commercial advertisements. Radio advertising was well on its way toward perfection a decade into Radio's "Golden Age." For example, Cantril and Allport analyzed radio advertising techniques in their 1935 book, *The Psychology of Radio*.[8]

It is easy to describe Radio (and later Television) commercials or the sponsors of American sports. This was mostly beer, sometimes of a

[8] See "The Visitor in Your Living Room: Radio Advertising in the 1930s" in American Studies at the University of Virginia Program Guide.

national brand, but very often local, popular in one region and unknown elsewhere, such as Narragansett Lager beer in New England, made famous by Boston Redsox baseball announcer Curt Gowdy. The underlying theme of the beer advertisements was refreshment, staying cool, relaxing and improving sociability, echoed by the phrase: "Hi, Neighbor! Have a 'Gansett.'" Tobacco advertisements, ranging from cigarettes to snuff and "plug" (chewing tobacco) played a role, also. Eventually, automobile manufacturers emerged as sponsors of college and professional sports broadcasted on the air waves.

The way in which many automobiles were advertised on radio during its "Golden Age" is interesting to recall. It was by sound, as much as by description. Of course, a narrator described the key features of a new car's (or truck's) design: its passenger payload, for example, or the amount of luggage that a car (baggage that a truck) could carry. In the background, almost subliminally, was the vehicle's own peculiar sound as the motor vehicle gained speed. As they gained speed, virtually all Ford vehicles of the era produced a sound which resembled the wind. Almost all General Motors (GM) vehicles made a "whining" sound that was heavier for the more expensive brands and lighter for the less expensive ones, ranging (top to bottom) from Cadillac, Oldsmobile, Buick, Pontiac, to Chevrolet. GM's line of trucks, particularly "GMC" and Chevrolet trucks, also had a peculiar "whine" as they moved across at least three manually shifted gears. Typically, GM automobiles retained this sound as they progressed from manual to automatic shifting. Since this was considered to be technological progress, these sounds became a part of radio commercials. Sometimes they were also incorporated into the backdrop of radio dramas.

We have already discussed the role which soap companies played in sponsoring afternoon situation comedies ("sitcoms") known as "Soap Operas." Variety shows also tended to receive sponsorship from the distributors of products bought by housewives, particularly food. For example, "The Kraft Music Hall" was sponsored by Kraft Corporation, distributor of cheese and other dairy foods. Heinz and Hellman's, then national and now global purveyors of ketchup, mayonnaise, mustard, and related condiments, sponsored similar shows. Towards the very end of Radio's "Golden Age," just before the rise of television, sponsors came to include producers of baby diapers as well as somc feminine hygiene products such as Kotex and Modess sanitary napkins. Condoms were never advertised prior to the Supreme Court's 1965 ruling that legalized their advertisement and sale in *Griswold v. Connecticut*.[9]

[9] U.S. Supreme Court. *Griswold v. Connecticut*.

General Electric Company sponsored "GE College Bowl Theater," a quiz show for college age contestants, with its slogan: "At General Electric, progress is our most important product." The same could be said for Radio generally. As a technological innovation, of course, the radio contributed to the development of broadcasting techniques. Intellectual progress on Radio came from its programs: from the fact that one broadcast medium could deliver comedy, drama, music, and sports (plus news and weather) via one piece of living room (or bedroom) furniture, present in almost every home. It can be argued that the progress which America made as a Nation during and immediately after the Depression was the movement from a collection of regions to a Nation State, largely because of the contribution which Radio made to a unified American culture.

Conclusion

American Radio programming was instrumental in creating the American Nationalism the country needed between 1926 and 1956, Radio's "Golden Age," as the Nation prepared to enter, wage, then recover from World War Two. This is almost entirely because Radio used as its centrifugal force what we term "Memorideas," or the collective glorification of concepts, events, and people from America's history in order to forge together a People true to the National Slogan: *e pluribus unum* ("One Out Of Many"). Comedy, Drama, Politics, Sports were blended together with Music, News, and the Weather to create "One America," without which the 20th century may not have become the American Century.

Works Cited

American Studies at the University of Virginia Program Guide. *America in the 1930's: On the Air*. Charlottesville, VA: University of Virginia, 2004. http://xroads.virginia.edu/~1930s/RADIO/radiofr.html (accessed March 19, 2009).

Mid-Hudson Regional Information Center (MHRIC). *Fireside Chats of Franklin D. Roosevelt*. http://www.mhrcc.org/fdr/fdr.html (accessed March 19, 2009).

Packer, Randall, and Ken Jordan, eds. "Pioneers." In *Multimedia: From Wagner to Virtual Reality*. New York: W. W. Norton, 2002. Also available online at http://www.artmuseum.net/w2vr/timeline/Wagner.html (accessed March 19, 2009).

Short, K. R. M., ed. *Western Broadcasting over the Iron Curtain*. New York: St. Martin's Press, 1986.
U.S. Supreme Court. *Griswold v. Connecticut*. 381 U.S. 479, 1965.

INTERPRETING THE NEW DEAL: THE PUBLIC OF THE FIRESIDE CHATS

RAFFAELE ASQUER

This article will explore one case of communication technology affecting American politics, namely the so-called fireside chats delivered by President Franklin D. Roosevelt during the New Deal period. In particular, attention will be focused on a well-known and controversial episode, that is, the fireside chat of March 9, 1937, dealing with the federal judiciary reform; in the second part, the public response to the radio broadcast will be analyzed in detail.[1] Drawing inspiration from recent trends in media and cultural studies as well as literary criticism, the present essay aims to illustrate how the New Deal and the figure of Franklin Roosevelt were apprehended and received by the "interpretive communities" that constituted the public of the fireside chats; also, it will provide an example of active media consumption in American political life.[2]

For our purposes, it may be useful to make some preliminary considerations. Firstly, it was neither Roosevelt nor his staff who labeled certain radio speeches as "fireside chats": it was the radio industry and the national press which did so. An informal and colloquial approach to the matter was supposed to be the distinctive feature of these broadcasts; also, they were devoted to a specific subject and were delivered purposely to address the problem, not for a particular occasion. Secondly, the fireside chats did not occur frequently at all. Actually, a large part of the success of the radio speeches is ascribed to their very rarity. They were "media

---

[1] This article is an excerpt from a dissertation written for the *Laurea Specialistica* (Second Degree) in Contemporary History at the University of Pisa (Italy): *"We, the People, Are With You": FDR, Huey P. Long and Their Public; Rhetorical Strategies and Political Identities.* The author would like to thank the staff of the Franklin D. Roosevelt Presidential Library, where he conducted his research, as well as the *Scuola Normale Superiore di Pisa* which made it possible through a series of research grants.

[2] Of course, I refer to Fish, *Is There a Text.*

events," as Michael Ryfe has argued, because they were so few: it is generally agreed that Franklin Roosevelt delivered 13 fireside chats during the New Deal period (March 1933–September 1939). Thirdly, after each transmission the president would receive thousands of letters from private citizens, as well as from trade union sections, cultural or religious associations. To a certain extent, it can be said that the presidential leadership of public opinion, which played a pivotal role for the entire Roosevelt administration, rested upon a sort of "revolution in mass political letter-writing," as pointed out by Leila Sussmann: both the dimension and the social composition of the letter-writing public were unprecedented.[3]

The Fireside Chat of March 9, 1937

More an appeal to the nation than a report, the fireside chat of March 9, 1937 was meant to help President Roosevelt in one of the most troubling circumstances of his twelve-year mandate. After his 1936 landslide, Democratic majority was firmer than ever and the path to further reforms seemed open, as the polls revealed the emergence of a New Deal social coalition; however, such reforms as the National Labor Relations Act still awaited to be confirmed by Supreme Court rulings, and the past record of the Court concerning New Deal legislation was deeply worrying for the president, who had expressed his impatience both on public occasions and in private meetings with members of the Cabinet. Although in his annual message to Congress (January 6, 1937) Roosevelt invited the Supreme Court to join him in the struggle against Depression, he was not willing to squander his extraordinary political capital in the process, and the great popularity which he enjoyed after November 1936 emboldened him to attempt a legislative solution. Unlike most New Deal legislation, however, the reform of the federal judiciary was mainly conceived by Franklin Roosevelt with the aid of Attorney General Homer Cummings, without consulting other members of the Cabinet or Congress; thus, when the bill was submitted to Congress on February 5, 1937, neither the political establishment nor public opinion were prepared. While the Judiciary Reorganization Bill contained many provisions, the most notorious one would have allowed the president to appoint an extra Supreme Court Justice for every sitting Justice over the age of seventy who refused to retire; thus, Roosevelt would have the possibility to appoint six more

[3] Ryfe, "Franklin Roosevelt"; Sussmann, *Dear FDR*.

Supreme Court justices immediately, increasing the size of the Court to fifteen members.[4]

As for the fireside chat, the opening was devoted to reminding the public of the worst days of the economic crisis, that is the very moment when President Roosevelt had started his mandate and his battle against the Depression as well; usually, that gloomy image was contrasted with the achievements of the New Deal, but here it was meant to be a disturbing caveat rather than a term of comparison. Surprisingly enough, given the usual reassuring tone of Roosevelt's broadcasts, a sense of emergency loomed large:

> We are at a crisis, a crisis in our ability to proceed with that protection. It is a quiet crisis. There are no lines of depositors outside closed banks. But to the farsighted it is far-reaching in its possibilities of injury to America.
>
> I want to talk with you very simply tonight about the need for present action in this crisis—the need to meet the unanswered challenge of one-third of a nation ill-nourished, ill-clad, ill-housed.[5]

In the following part, the speech suggested an opposition between popular will and the Court: here, the president clearly acted as a "leader-interpreter" *a la Wilson*, claiming to speak for the American people.[6] He repeated the metaphor which he had used in his Democratic Victory Dinner address five days before and described the US form of government as "a three horse team provided by the Constitution to the American people so that their field might be plowed"; consequently, the current situation whereby "Two of the horses are pulling in unison today, the third is not" implied a departure from the direction that had already been taken by the American people, meant as the founding subject of the democratic system. Actually, this turned out to be a fallacy in the president's

---

[4] It should be noted that, since no constitutional provision specified the size of the Supreme Court, the number of judges had varied from six (1789) to nine (1869), with a maximum of ten (1863), by ordinary legislation.
On the Judiciary Reorganization Bill of 1937, its background and consequences, as well as later developments of the Supreme Court, leading to a series of pro-New Deal rulings, see Badger, *New Deal*, 263–7 and Leuchtenburg, *Franklin D. Roosevelt and the New Deal*, 213–20.

[5] Buhite and Levy, *FDR's Fireside Chats*, 86. The text presented in this article is based on the transcription of the radio recordings; on the divergence between the radio recording and the official version of the fireside chats, see ibid., xviii–xix.

[6] For the concept of "leader-interpreter" in Woodrow Wilson's doctrine of presidential leadership, see Tulis, *Rhetorical Presidency*, 125–37.

argument, because the power of the Supreme Court did not depend on popular will (nor does it); rather, it rested upon the Constitution itself, which had enshrined it as the highest judicial body. Though a questionable move, equating the three branches of government led to compelling political consequences, which Roosevelt moved on to illustrate: first, the American people had proved willing to follow the president, conferring upon him and the Congress a mandate to protect the country from the Depression; then, the Court had prevented the executive and legislative branches from meeting the need, since it had began to act as "a third house of the Congress—a super-legislature"; eventually, the American people—through Roosevelt's voice, of course—claimed the prerogative to step in and recover political agency.[7]

> It is the American people themselves who are in the driver's seat.
> It is the American people themselves who want the furrow plowed.
> It is the American people themselves who expect the third horse to pull in unison with the other two.[8]

Not only was the Court acting against popular will, but also against the very Constitution which it was supposed to protect. The Constitution, as the president explained, gave birth to a stronger national government than the Articles of Confederation, so that it could "provide for the common defense" and "promote the general welfare" of the United States; it being impossible to forecast future needs, the Congress was given all the powers required to meet each and every problem which had a national character. Recently, though, the Supreme Court had thwarted congressional efforts to tackle current needs, clearly national in character, such as "to stabilize national agriculture, to improve the conditions of labor, to safeguard business against unfair competition"; it was misinterpreting not only what the American people asked for, but also its own role in the constitutional edifice. In order to support his view, President Roosevelt invoked the authority of dissenting Supreme Court judges, with reference to recent cases that had involved New Deal legislation.

---

[7] Interestingly enough, Franklin Roosevelt seemed to endorse a "strict" philosophy of constitutional interpretation that is the very approach that liberals would oppose in the following decades; conservatives, instead, would denounce the attempts by the Supreme Court to act as "a third house of the Congress."

[8] Buhite and Levy, *FDR's Fireside Chats*, 86–87.

> In the face of such dissenting opinions, there is no basis for the claim made by some members of the Court that something in the Constitution has compelled them regretfully to thwart the will of the people. . . .
>
> We have, therefore, reached the point as a nation where we must take action to save the Constitution from the Court and the Court from itself. We must find a way to take an appeal from the Supreme Court to the Constitution itself. We want a Supreme Court which will do justice under the Constitution and not over it. In our courts we want a government of laws and not of men.[9]

Given these premises, Roosevelt offered his solution to the audience, that is, a reform through legislation, and made his case against the alternative solution that was under discussion, which was a constitutional amendment. First, he explained the twofold rationale underlying his proposal to confer upon the president more influence in the Supreme Court composition; the plan would bring into the judicial system "a steady and continuing stream of new and younger blood," so as to make the administration of federal justice speedier and less expensive; in addition, newly-appointed judges would represent "personal experience and contact with modern facts and circumstances under which average men have to live and work," that is a sharper awareness of the economic and social problems facing the nation. Second, Roosevelt collected a series of arguments supporting his case: for example, the statute would apply to all the courts in the federal system—a provision which had been greeted with almost universal approval; why were the same principles—an efficient and up-to-date judiciary—so bitterly contested when applied to the Supreme Court? Besides, he recalled the fact that the number of justices had been changed several times before, for instance by such revered presidents as John Adams and Thomas Jefferson, so it was not a long-standing tradition in American polity. Finally, Roosevelt discussed a possible constitutional amendment and concluded that it would take years to agree on a single draft, get a two-thirds majority in both Houses of Congress and get through the ratification process by three-fourths of all the States; against the background of a "quiet crisis," of course, no delay could be accepted by the American people, who had learnt the importance of quick action.

> Like all lawyers, like all Americans, I regret the necessity of this controversy. But the welfare of the United States, and indeed of the Constitution itself, is what we all must think about first. Our difficulty with the Court today rises not from the Court as an institution but from human beings within it. But we cannot yield our constitutional destiny to the

[9] Ibid., 89.

> personal judgment of a few men who, being fearful of the future, would deny us the necessary means of dealing with the present.
>
> This plan of mine is no attack on the Court; it seeks to restore the Court to its rightful and historic place in our constitutional government and to have it resume its high task of building anew on the Constitution "a system of living law." The Court itself can best undo what the Court has done.[10]

After trying to soothe the heated controversy over his proposal, the president addressed those who opposed his plan on the ground that they favored a constitutional amendment; here, his political strategy aimed to win over those Democratic supporters or pro-New Deal fractions of public opinion that were increasingly cold if not opposed to the scheme. Thus, he returned to the "economic royalists'" argument, so to speak, which had proved such an effective rhetorical weapon in the recent election campaign: after mentioning the opposition by "a very large percentage of newspaper publishers, Chambers of Commerce, Bar Association, Manufacturers' Associations," he contrasted their economic interests with the will of the American people, as expressed at the ballot box. Obviously, no clear mandate on the issue had emerged from the 1936 elections, but the argument was meant to provoke a prompt reaction: if forced to choose between the American people and the "privileged few," an anonymous listener would probably take the people's side—that is, President Roosevelt's.

> The first [group] includes those who fundamentally object to social and economic legislation along modern lines. This is the same group who during the campaign tried to block the mandate of the people.
>
> And the strategy of that last stand is to suggest the time-consuming process of amendment in order to kill off by delay the legislation demanded by the mandate.
>
> To those people I say, I do not think you will be able long to fool the American people as to your purposes. . . .
>
> The present attempt by those opposed to progress to play upon the fears of danger to personal liberty brings again to mind that crude and cruel strategy tried by the same opposition to frighten the workers of America in a pay-envelope propaganda against the Social Security law. The workers were not fooled by that propaganda then. And the people of America will not be fooled by such propaganda now.[11]

---

[10] Buhite and Levy, *FDR's Fireside Chats*, 93.
[11] Buhite and Levy, *FDR's Fireside Chats*, 94–95.

Lastly, after restating his reasons briefly, the president reaffirmed the purpose to restore the balance between the three branches of government, which was threatened by the Court; thus, he tried to deflect the deluge of accusations leveled at the federal judiciary itself, so as to present himself as the savior of the Constitution as well as the interpreter of the American people.

> You who know me will accept my solemn assurance that in a world in which democracy is under attack, I seek to make American democracy succeed. You and I will do our part.[12]

The Public Response

As far as the public response to the fireside chat is concerned, the federal judiciary reform can be read as a debate over different conceptions of "the people": indeed, this was the key term on which the reaction of the letter-writing public was centered. Moreover, as the reform eventually failed, it may be interesting to explore in detail both the favorable and unfavorable side of the president's correspondence.[13]

As for the favorable letters, they tended to frame their support in terms of social welfare and economic security, which were usually referred to "the people" meant as population. As in the response to the fireside chat of September 30, 1934, the underlying rationale for this position rested upon an assessment of the New Deal and President Roosevelt as a program and a leader committed to aid "the masses of the people" or "the common people" of the country; at least for sympathetic letter-writers, such a representation still shaped a large part of the correspondence, probably as a result of the political and rhetorical shift carried out by the

[12] Ibid., 95.

[13] All the letters quoted hereinafter are taken from President's Official File 41 (OF 41), FDRL. Correspondence written in reaction to the fireside chats is currently housed in the Franklin D. Roosevelt Library, where letters are usually arranged by the date of the broadcast to which they responded, the names of the respondents in alphabetical order, and their positive or negative attitude. However, the letters in reply to the fireside chat of March 9, 1937 are not arranged separately from the general correspondence of the same period. Thus, in order to single out the correspondence written in response to that radio speech, the following method was adopted: among the thousands of letters to be found in the President's Official File of those months, those written in the first week after the fireside chat were selected for this research, as well as those which contained some reference to the speech. At the end of this process, a sample of 440 items was pulled out randomly to be examined in detail.

administration from 1935 on. Within this framework, the Supreme Court rulings on New Deal legislation were interpreted as a direct assault on "the welfare of the people," which was of paramount concern for this part of the letter-writing public; given the distressing circumstances of the Depression, the needs of "the people" were to be met as quickly as possible, so the only solution was to pass the judiciary reform through ordinary legislation. Interestingly enough, the same argument could involve "the people" as a political subject as well, "the people" meant as the founding subject of the political system and democratic discourse: in 1936, "the people" had voted for Franklin Roosevelt and the New Deal, thus demanding economic security for "the people," here taken to mean simply "the population." Incidentally, President Roosevelt's fireside chat had moved along the same lines, combining the principle of popular rule and the need for social protection: "The American people have learned from the depression. For in the last three national elections an overwhelming majority of them voted a mandate that the Congress and the president begin the task of providing that protection—not after long years of debate, but now."[14]

> It is time to institute such reforms as is necessary for the welfare of the country. You have received a mandate from the people to go ahead. Do not let them deter you from your path.[15]

> Since it is your earnest desire to put into effect certain policies and reforms for the betterment and the uplift of the masses of the American people, and since the Supreme Court of the United States stands as a barrier for such legislation, it is my opinion that every right thinking man or woman in this country will consent that it is high time a change should be made.[16]

> The opposition to your Court reorganization plan, as well as to any other plans designed to benefit the common people and promote their general welfare, comes from a class of people who are, for the most part,

---

[14] Buhite and Levy, *FDR's Fireside Chats*, 85. "Protection—writes Maurizio Vaudagna—is a word resonant with the ambiguities of paternalism and paternity, which imply that the protected subjects are lovingly taken care of by their benefactor but are nevertheless also hierarchically subject to and dependent on him" (Vaudagna, "Victorian Virility," 583). This remark, in the context of the "moralization of modernity" expressed by the fireside chats, should be kept in mind whenever "protection" is given as a rationale for New Deal reforms.

[15] Anonymous to FDR, no date, Newark (New Jersey), OF 41, Box # 11, FDRL.

[16] A. Bayer to FDR, March 11, 1937, Olympia (Washington), OF 41, Box # 3, FDRL.

> political "yes men," men who would not even be holding a political job if it were not for the fact that your name headed the ballot on which they ran.
> We, the common people, are with you![17]

"The common people" proved an evocative yet ambiguous expression, whose implications were difficult to handle. On the one hand, it could be used to characterize the New Deal and express the need for social protection, as required by a large part of the American population. On the other hand, it could be meant as part of an "opposition structure": that is, "the people" or "the common people" were contrasted with "the privileged few" or "a small minority," acting as terms of opposition; thus, the letters evoked the sense of an ongoing conflict between "the people" (and the president, of course) and a poorly defined and somewhat blurred enemy. Of course, this was a response to the confrontational rhetoric Roosevelt had been adopting since the 1936 election campaign, when he presented himself as the champion of the "average man" against the "economic royalists"—an argument he used to reiterate when he needed to regain political initiative. However, contrasting "the common people" with "the few" (broadly construed, of course) represented a secondary strategy in the response from the public: for most letter-writers, the New Deal was simply aimed to assist "the common people" and would continue to do so, with no reference whatsoever to such powerful opposition. Here, Roosevelt's confrontational rhetoric turned out to be ineffective, although his fireside chat had tried to remind the public of an ever-lasting struggle: "those who fundamentally object to social and economic legislation along modern lines. This is the same group who during the recent campaign tried to block the mandate of the people." Instead, whether directly inspired by the radio message or not, some letter-writers did interpret the judiciary reform issue as yet another phase in the struggle between the "privileged few" and "the people/common people," under the guidance of President Roosevelt.

> And I just want to say that the same forces who were against you in Nov. the Chamber of Commerce etc are the ones who are still against you, but we the common people who can't afford to send telegrams, are still for you and always will be . . .
> In closing I ask you to pay no attention to the wires and letters from the reactionaries protesting your Supreme Court just remember that they have nothing else to do but set around and frame up letters . . . while the workers

[17] Allen Baker to FDR, March 10, 1937, Cleveland (Ohio), OF 41, Box # 3, FDRL.

all tho [*sic*] 100% for you come home tired at nite and neglect to voice their feelings. but we do vote and we vote right.[18]

God bless and help you, Mr. President, for you have a hard fight on your hands, but don't be deceived by the old gang that backed Hoover, Landon and now the reactionaries on the Supreme Court; they yelled, controlled the press, sent telegrams etc. then, like now; but the silent, everyday, peaceful and democratic loving Americans are still with their champion in this fight, *Franklin Delano Roosevelt*![19]

As for the unfavorable letters, the rhetorical strategy underlying most of them was centered around "the people" as a political subject, whose political agency was under assault by the president. A close analysis of the correspondence makes it clear why these letter-writers appealed to the sovereign will of the people, and how the appeal was framed. First, it should be remarked that economic security and social protection were not of paramount concern, rather the focus was on the principle of popular rule and the need to protect it against a supposed violation by the president: "the people" acted as an ideal yardstick to measure the legitimacy of the current government, in particular of Roosevelt's legislative agenda. Sometimes, the appeal to "the people" became even more pressing when "the rights of the people" were called upon; political rights, indeed, were at stake since President Roosevelt had raised the issue of the Court reform and, most of all, had ruled out constitutional amendment as a possible solution—a solution through which "the voice of the people" would be heard again.

Last but not least—AND THIS IS THE MOST IMPORTANT OF ALL—you received no mandate from me last November to do anything of the sort and I remind you that the members of the Congress and yourself are but our agents—and that despite your exalted position, we are still the principals in this government and the only ones whose voice can be heard in altering our fundamental structure of government.[20]

---

[18] G. Atkinson, C. Atkinson, and S. Peterson to FDR, March 10, 1937, Los Angeles (California), OF 41, Box # 2, FDRL. Incidentally, it should be remarked that such appeal to a sort of "silent majority" was a common argument among the favorable letter-writers.

[19] Louis D. Lindo to FDR, March 9, 1937, Barre (Vermont), OF 41, Box # 10, FDRL.

[20] Francis J. Duffy to FDR, March 12, 1937, Yonkers (New York), OF 41, Box # 39, FDRL.

> You have done many good things while in office, but this time I pray God to defeat your plan which may wreck the Constitution, cast the Supreme Court into total disrespect, and take from the people a sacred right.[21]

> I would favor an amendment on which we, "the people," would have time to reflect and debate.[22]

As a result, Roosevelt's rhetorical and political strategy was rejected as a whole, with far-reaching implications for the success of the rhetorical presidency.[23] Whereas Roosevelt—and the favorable letter-writers as well—tended to conflate the different meanings of "the people," the opposition to the Court reform rested upon an accurate separation among the semantic areas the term referred to: even if "the welfare of the people" or "the people" in need of social protection were still present, surely they were not to be confused with "the people" as the political subject *par excellence*. Far from being cast aside by the "leader-interpreter," "the people" claimed their supremacy and tried to recover their political agency: "None of us should fear the will of the people when expressed directly on any issue. The people have not expressed themselves on the issue of the Supreme Court. If we must do this job, it should be done by a Constitutional amendment."[24] As for "the common people *against* the few," Roosevelt's argument was disproved as well: realizing how powerful such a rhetorical weapon could be, some writers proved eager to articulate the position from which they were speaking; thus, they exposed the weakness of the opposition structure envisioned by the president and shared by his supporters.

> Just read your speech STOP Please stop saying "we" in referring to meddling with Supreme Court STOP All Court fixers eventually go to jail

---

[21] An Independent Voter [pseud.] to FDR, March 11, 1937, no address, OF 41, Box # 41, FDRL.

[22] H. Books to FDR, March 10, 1937, Pittsburgh (Pennsylvania), OF 41, Box # 39, FDRL.

[23] Of course, I refer to Tulis, *Rhetorical Presidency*. See also Ceaser et al., "Rise of the Rhetorical Presidency," 158–71. For critical debate on the rhetorical presidency and related subjects, see Ellis, *Speaking to the People*. For Franklin Roosevelt's case in particular, see Ryan, *Franklin D. Roosevelt's Rhetorical Presidency*.

[24] R. Andersen to FDR, March 10, 1937, Auburn (New York), OF 41, Box # 39, FDRL.

(you're not smart enough to get by) and I don't want to be with you when you go.[25]

> Please, Mr. President, give us credit for not being in the class you refer to as "Economic Royalists," and as being as sincere and patriotic in our opinions as those who agree with you.[26]

Clearly enough, this part of the letter-writing public reacted to the "appropriation of the people" which lay at the core of Franklin Roosevelt's rhetorical presidency: though able to "go over the heads" of Congress, the president could not always rely on a receptive public.

Conclusion

In conclusion, the fireside chat of March 9, 1937 can be considered a striking example of both the resources and the limits of the rhetorical presidency. On the one hand, the president could enjoy a valuable opportunity to influence public opinion on such a controversial subject, without being directly challenged by political opponents. It was Roosevelt and his closest aides who decided to tackle the problem and came up with the solution. Congress was accorded scant consideration, but it soon became clear that not even the Democratic majority was solid and sympathetic enough to pass another bill coming from the White House; now, more than ever, the president had to rely on the American population to find support for his agenda. On the other hand, this strategy exposed the president to a political backfire, mainly as a result of his overconfidence. Firstly, facing such an uncompromising approach, the Congress rejected the proposal and, even more importantly, conservative Democrats joined Republicans in their first anti-New Deal campaign. Secondly, public opinion itself seemed unready to support the Court reform, either concerned about the checks and balances system or fearing an excessive concentration of power in Roosevelt's hands. In any case, through the Gallup polls as well as the correspondence to the White House, a vocal opposition came to the surface, making it easier for representatives to resist the assault of the rhetorical presidency.[27]

---

[25] Abe Bloom to FDR, telegram, March 10, 1937, no address, OF 41, Box # 39, FDRL.

[26] Arthur Boreman to FDR, March 10, 1937, Des Moines (Iowa), OF 41, Box # 39, FDRL.

[27] Levine and Levine, *People and the President*, 167.

Works Cited

Primary Sources

Buhite, Russell D., and David W. Levy, eds. *FDR's Fireside Chats*. New York: Penguin Books, 1993.

Franklin D. Roosevelt Presidential Library (FDRL). Hyde Park, New York. Official File 41 (OF 41).

Secondary Sources

Badger, Anthony. *The New Deal: The Depression Years, 1933–1940*. New York: Hill and Wang, 1989.

Brandenburg, Earnest. "The Preparation of Franklin D. Roosevelt's Speeches." *Quarterly Journal of Speech* 35 (1949): 214–26.

Brandenburg, Earnest, and Waldo W. Braden. "Franklin D. Roosevelt's Voice and Pronunciation." *Quarterly Journal of Speech* 38 (1952): 23–41.

—. "Roosevelt's Fireside Chats."*Speech Monographs* 22 (1955): 291–302.

Ceaser, James W., et al. "The Rise of the Rhetorical Presidency." *Presidential Studies Quarterly* 11 (Spring 1981): 158–71.

Craig, Douglas B. *Fireside Politics: Radio and Political Culture in the United States, 1920–1940*. Baltimore: John Hopkins University Press, 2000.

Ellis, Richard J., ed. *Speaking to the People: The Rhetorical Presidency in Historical Perspective*. Amherst: University of Massachusetts Press, 1998.

Fish, Stanley. *Is There a Text in This Class? The Authority of Interpretive Communities*. Cambridge, MA: Harvard University Press, 1980.

Leuchtenburg, William E. *Franklin D. Roosevelt and the New Deal*. New York: Harper and Row, 1963.

Levine, Lawrence L., and Cornelia R. Levine. *The People and the President: America's Conversation with FDR*. Boston: Beacon Press, 2002.

Lim, Elvin T. "The Lion and the Lamb: De-mythologizing Franklin Roosevelt's Fireside Chats." *Rhetoric and Public Affairs* 6 (2003): 437-64.

Markowitz, Gerald, and David Rosner. *Slaves of the Depression: Workers' Letters about Life on the Job*. Ithaca and London: Cornell University Press, 1987.

McElvaine, Robert S. *Down and Out in the Great Depression: Letters from the "Forgotten Man."* Chapel Hill: University of North Carolina Press, 1983.

Ryan, Halford R. *Franklin D. Roosevelt's Rhetorical Presidency.* Westport, CT: Greenwood Press, 1988.

Ryfe, David M. "Franklin Roosevelt and the Fireside Chats." *Journal of Communication* 49 (Fall 1999): 80–103.

—. "From Media Audience to Media Public: A Study of Letters Written in Reaction to FDR's Fireside Chats." *Media, Culture and Society* 23 (November 2001): 767–81.

Schmidt, William T. "Letters to Their President: Mississippians to Franklin D. Roosevelt, 1923–1933." *Journal of Mississippi History* 40, no. 3 (1978): 231–52.

Steele, Richard W. *Propaganda in an Open Society: The Roosevelt Administration and the Media, 1933–1941*. Westport, CT: Greenwood Press, 1981.

Sussmann, Leila. *Dear FDR: A Study of Political Letter-Writing*. Totowa, NJ: Bedminster Press, 1963.

Tulis, Jeffrey K. *The Rhetorical Presidency*. Princeton, NJ: Princeton University Press, 1987.

Vaudagna, Maurizio. "Victorian Virility, Democratic Emotionalism and Patriotic Citizenship in Franklin D. Roosevelt's Fireside Chats." In *Public and Private in American History: State, Family, Subjectivity in the Twentieth Century*, edited by Raffaella Baritono and others. Torino: Otto Editore, 2003.

Winfield, Betty Houchin. *FDR and the News Media*. New York: Columbia University Press, 1994.

IS ANYBODY LISTENING? VICE PRESIDENTIAL CANDIDATES AND THEIR TV DEBATES

MACIEJ TUREK

For the major part of U.S. history, the vice presidential office was perceived as one of the least meaningful, if not useless, positions in the federal administration. In the words of John Adams, the first American vice president, it was "the most insignificant office that ever the Invention of Man contrived or his Imagination conceived."[1] However, political life treated vice presidents even worse, especially after the enactment of the Twelfth Amendment. The office was rather absent from public awareness: few took its holders seriously, and even fewer thought of running for vice presidency.

Over the last few decades, however, the significance of the office has grown significantly. Vice presidents have gained more power, played more important political roles, their position becoming a full-time job. It is unlikely today that an open conflict between the President and No. 2 could occur, as it did between John Adams and Thomas Jefferson, or Andrew Jackson and John C. Calhoun. It is hard to imagine that while holding the office, a vice president would run a tavern, as did Richard Johnson, vice president in the Martin Van Buren administration (1837–1841). It is not possible that during a great conflict, with the United States politically and militarily involved, the only two elective officers rarely speak to each other, as was the case with FDR and Harry Truman. Finally, it seems improbable that in the future, as it often happened in the past, the activity of the running mate will end with the result of the Election Night.

Contemporary vice presidents are much more important players, with real power and authority. It is true that what the vice president will do or is permitted to do is wholly determined by what the President assigns or allows him.[2] Nevertheless, Presidents also recognize that there can be

---

[1] Diggins, *John Adams*, 42.

[2] Humphrey, *Changes in the Vice Presidency*, 58.

more use of a vice president who is an active spokesperson of the administration than of one remaining in the shadow. Over the past decades "Veeps"[3] have indeed risen in status, becoming special envoys, commission chairmen, top presidential advisers, administration mediators and lobbyists in Congress, presidential and party liaisons, goodwill ambassadors and valuable campaigners. The last role—vice presidents and running mates as campaigners—turns out to be one of high importance.

I

Campaigning skills became a necessity as the status of the second men—or, more precisely, second persons—increased. Of course, the first vice presidents engaged in campaigning, in the fight for the presidency. According to the U.S. Constitution, the person who came second in the presidential race automatically became the vice president. This regulation was changed with the Twelfth Amendment that introduced a separate vote for the first and the second office. However, this rule made vice presidency quite an unpopular post to run for—a political dead end. Few ambitious politicians would strive for it. If they accepted the nomination, there was the traditional main goal for them to achieve in the campaign—to carry their home state, usually one with a large number of electoral votes to gain. The first vice presidential candidate to run a broad campaign was Theodore Roosevelt, running on the ticket with President William McKinley in 1900. According to Mark O. Hatfield, Roosevelt's campaign dominated the news:

> . . . President McKinley conducted his "front porch campaign," just as he had in 1896, except this time he received guests at the White House rather than his home in Canton, Ohio. This strategy suited the vigorous Roosevelt extremely well, as he proclaimed himself to be "strong as a bull moose." It allowed him to tour the West and Midwest, taking on Democratic presidential candidate William Jennings Bryan on issues of the tariff, the gold standard, and American empire. These two great orators set standards of stamina never before seen. Roosevelt covered 21,000 miles in twenty-four states, making over 600 speeches. Roosevelt's tour helped the GOP . . . in the West and added life to an otherwise dull campaign. The vice-presidential candidate radiated energy, while McKinley sat on his porch in Washington, reminding the nation how prosperous it was.[4]

---

[3] The term "Veep" is a colloquialism for vice president. It is in use since Alben Barkley's vice presidency (1949–1953).

[4] Hatfield, *Vice Presidents of the United States*, 301–2.

The next second man to campaign so actively was Richard Nixon, who, however, was running in very different circumstances. On April 12, 1945, a four-time presidential election winner Franklin Delano Roosevelt died, and the presidency was assumed by Harry S. Truman, who in the opinion of most historians, was unprepared to lead the country at the crucial time, which was the final stages of World War II. Though years after his term Truman is rated as one of the best American presidents, this was not yet so clear in the late 40s and early 50s. To prevent the country from being led by another "accidental president," a new approach to selecting a running mate was set, and vice presidential candidates were chosen with more attention. This makes vice presidency more attractive for ambitious, high-caliber politicians, but also requires more activity from the candidate.

Nixon's position was also special, because he was running with General Dwight Eisenhower. As a wartime hero and a former NATO Supreme Commander, Ike was not an active politician. As late as a year before the election, he announced he would accept the nomination from the Republican Party, despite having also been offered a standard bearer position by Democrats. Thus, Eisenhower, with no party background, needed the young and lively Nixon, then a rising star of the GOP, not only to conduct a campaign, but also to promote Eisenhower as a candidate and at the same time prevent a split inside the party. This was to become the basic role for future running mates—while the presidential nominees pitch their appeals to potential switchers, the vice presidential candidates must maintain the loyalty of party regulars.[5] The role of mobilizing party support is one of the more crucial jobs which the running mates fill.[6]

Nixon also set another precedent. Eisenhower was determined to be president of "all the people,"[7] and did not want to be involved in party politics. Therefore, Nixon was put in charge of the 1954 mid-term Republican campaign. He traveled over 26,000 miles, visited 95 cities in 31 states, delivered 204 speeches, and met the press over 100 times.[8]

From that time on, Veeps have been playing a new role in the campaign strategy. As vice presidential scholar Joel Goldstein notices, "The vice-presidential candidate, like the campaign manager, press aide, speechwriter, or treasurer, is part of the campaign team."[9] However, it

---

[5] Goldstein, *Modern American Vice Presidency*, 83.
[6] Ibid., 95.
[7] Hatfield, *Vice Presidents*, http://www.senate.gov/artandhistory/history/resources /pdf /richard_nixon.pdf.
[8] Young, *American Roulette*, 262.
[9] Goldstein, *Modern American Vice Presidency*, 91.

seems that running mates are more than just a part. Their participation, as broad as possible, could be beneficial not only for the standard bearers, but also for the second persons themselves.

The usefulness of vice presidential campaigning for No. 1 men is obvious: 1) Presidential candidates cannot go everywhere. The running mate can cover places the number-one man does not reach; 2) as Goldstein observes, "[p]residential candidates cannot sing their own praises publicly to the extent they might like. Their running mates need show no such inhibitions"[10]; 3) finally, "[p]ersonal criticism is an entrenched aspect of American presidential campaigns." But nasty abuse of opponents by the standard-bearer can discourage potential voters. Thus all the "dirty work" is left for second men, who "are often responsible for making the most intense attacks."[11]

For instance, Lyndon Johnson's "ability, as a party organizer, defender of Kennedy and hit man, especially in the South, was crucial to the Democratic victory in 1960."[12] Johnson's 23-year presence and high-level visibility in politics, connected with his Texas origins, left Kennedy free to undertake an intensive campaign in the rest of the country. In a close race, such as the 1960 election, it can be decisive for the outcome. On the other hand, as Light observes, vice-presidential campaigning brings benefits to running mates themselves: "Campaigning not only exposes the vice president as a national figure, but allows the office holder to make contacts that can be used in the future."[13]

> [N]othing helps a vice president look quite as presidential as campaigning, speaking and public liaison. . . . Since vice presidents cannot make decisions that the public can see, the political role [such as that of a campaigner] becomes the best vehicle for gaining name recognition and visibility.[14]

JFK's vice presidential bid of 1956 was in fact a maneuver to appear in the minds of Democratic politicians and voters as a possible, future candidate for a presidential nomination. It was rather unlikely that, even if it occurred, the Stevenson-Kennedy ticket could beat "Ike and Dick."

Nixon's activities as a campaigner, and, in fact, as the first vice president to become the party leader, gave him many supporters inside the

---

[10] Ibid., 100.
[11] Ibid., 111.
[12] Ibid., 112.
[13] Light, *Vice-Presidential Power*, 114.
[14] Ibid., 117.

GOP. This was certainly a highly significant factor in the recovery of Richard Nixon's political career. After two lost campaigns—for the White House in 1960 and for the governor of California in 1962—no one was going "to have Nixon to kick around anymore."[15] Even though he had not participated in "big politics" for a couple of years, the former vice president was able to receive the 1968 Republican Party presidential nomination. A similar situation occurred in 1984, when former vice president Walter Mondale, also a great campaigner, became the Democratic Party presidential nominee.

Going through Nixon's memoirs, as well as biographies and monographs, one may conclude that the same factors went into the rise and fall of this politician. Without them vice president Nixon would never have become a superstar; yet without them there would also have been no Watergate exposure. Those factors were the journalists and the media.

II

Certainly, the media impact does not relate only to Richard Nixon, but to every single politician, whether on the local or national level. In many aspects, the journalists' involvement, as well as the increasing amount of live coverage changed the course of history.

With the first radio broadcasts early in the 1900s, it could be expected that sooner or later the media would become interested in politics. Indeed, one of the first live national broadcasts was the inauguration of Calvin Coolidge in 1923. But the real *maestro* in using the media as a political tool was president Franklin Delano Roosevelt. It was on the radio that he presented the New Deal assumptions and his vision of the economic recovery. With a series of radio talks, the so-called fireside chats, FDR was able to convince Americans to trust the government's policy. In addition, due to radio shows and meetings with the press corps every week, Roosevelt built his position as an efficient politician. After FDR, it became standard practice that Americans learned about policy directions or important national and global political developments on the radio.

However, the real revolution started with the expansion of the television networks and mass production of TV sets after World War II—in 1945 only the richest people could afford to have a TV set. By 1962, nine out of ten American families had one. With the media being the new reality in the United States, it must have changed the style of the political campaigns. Television brings politicians right into the living room and lets

---

[15] Ambrose, *Nixon*, 671.

voters form their own impressions instead of having to depend on what local party bosses, union leaders, church spokesmen, or business chiefs say. The impact of television on campaigns is undeniable. For many candidates it has replaced going door to door.[16]

The media became kingmakers[17]—the image they (TV stations in particular) create can help or thwart a candidate's hopes. The power of television was shown in the 1960 presidential election, when John F. Kennedy and Richard Nixon were running. It was during this election that a television debate between two candidates was held for the first time.

Political debating in the United States has had a long history—contests in debating are organized for high school students. Candidates for local and state offices, as well as politicians running for Congress debate with one another. As long ago as 1858, Abraham Lincoln and Stephen A. Douglas confronted each other in a series of seven Illinois Senate Debates. The format of those debates, however, would probably be absolutely unacceptable for any TV station today. Each of the Lincoln-Douglas clashes lasted three hours; the first candidate spoke for one hour, the second for an hour and a half, and then the first replied for a half hour.[18] In 1948 the first radio debate was organized, when Thomas Dewey and Harold Stassen were campaigning in a presidential primary in Oregon. Still, the only topic of the debate was outlawing the Communist party in the United States.[19]

Yet never before had the candidates for the highest office had to discuss the future of the country in a national broadcast. The *spiritus movens* of the presidential debates on TV was Lar Daly,

> a perennial candidate for one office or another, who had the habit of campaigning in a red-white-and-blue Uncle Sam uniform. In his 1959 campaign for a mayor of Chicago, he filed a complaint with the Federal Communications Commission because certain television stations had denied him equal time after showing film clips of other candidates, [which at that time was guaranteed by law at that time] . . . In 1960, [to broaden those rules], Congress opened hearings on whether television stations and networks should be required to make free time available to qualified presidential candidates.[20]

---

[16] Smith, *Power Game*, 36.

[17] See chapter "Elections in the Television Age," in Graber, *Mass Media*, 155–93.

[18] Commission on Presidential Debates (CPD), 1858 Debates, http://www.debates .org/pages/his_1858.html (accessed March 29, 2009).

[19] CPD, 1948 Debate, http://www.debates.org/pages/his_1948.html (accessed March 29, 2009).

[20] Windt, "1960 Kennedy-Nixon Debate," 2.

Though for Kennedy there was more to gain than to lose in the debates, and the opposite was true for Nixon, the Republican candidate disregarded his aides, who had advised him not to debate on TV. Nixon decided to debate with Kennedy for two reasons. Firstly, he claimed, because the election would undoubtedly be close, and the debates might be crucial for winning. Secondly, Nixon had made his career out of successful debating experience and skillful use of television, not the least of which was his famous "Checkers" speech in 1952, and his "kitchen debate" with Russian leader Nikita Khrushchev in 1959.[21] In fact, the set of four debates (or at least the first) turned out to be disastrous for Nixon. For many viewers—and the number of viewers watching the first debate coverage was estimated at around 66 million[22] (in 1960 the U.S. population was around 180 million)—the young, vital, fresh and handsome Kennedy overshadowed the tired and grey-looking, five-o'clock-shadow-bearded Nixon. Because the debate occurred early in the campaign, was witnessed by the largest audience, and featured Kennedy's best performance, it shaped the rest of the 1960 campaign.[23] Kennedy managed to show his leadership qualities by altering the public impression that he was too young and inexperienced to be President.[24] The Schwerin Research Corporation reported in its viewer survey that Kennedy had outscored Nixon by 39 to 23 percent. The Gallup Poll showed Kennedy winning by 43 percent to 23 percent. Significantly, those who only heard the debate on the radio judged it to be much closer, a draw between the two.[25] The magic of TV was shown in a full range. If one considers that in the 1960 elections 14,000 votes in five states would have produced a different outcome, it is clear that the debates were one of the crucial events to change the course of history. Thus, the 1960 election marks an important turning point in the role of television in presidential election history.[26] Politicians were so awed by the risks of television debates that they shied away from a repeat performance for 16 years.[27]

When it was agreed that debates should be resumed, it made the news once more, and again made an impact on the election outcome. In the second 1976 presidential debate, President Gerald Ford asserted that "there

---

[21] Ibid., 3.

[22] All viewership data provided by Nielsen Media Research; see CPD, 1960 Debates, http://www.debates.org/pages/his_1960.html (accessed March 29, 2009).

[23] Hinck, *Enacting the Presidency*, 17.

[24] Graber, *Mass Media*, 160.

[25] Windt, "1960 Kennedy-Nixon," 20.

[26] Dover, *Presidential Elections in the Television Age*, 10.

[27] Graber, *Mass Media*, 160.

is no Soviet domination of Eastern Europe, and there never will be under a Ford administration."[28] The obvious *gaffe* was exposed by the media and the Democrats ever after in the campaign. However, the 1976 election was also important because of another debate—for the first time, the vice presidential candidates were able to present themselves in a nationally broadcast discussion.

III

The events from the past decades have shown that it actually could, and should matter who holds the second post. The sudden shock of President Roosevelt's death in the closing months of World War II awakened the nation as never before to the vital importance of an experienced vice president.[29] The events from the subsequent decades—the temporary running of the government by vice president Richard Nixon on the occasions of President Eisenhower's heart attack in September 1955, his ileitis in June 1956, and his stroke in November 1957; Lyndon Johnson's and Gerald Ford's successions, due to the JFK assassination and Richard Nixon's resignation respectively—reaffirmed the views that it was necessary to have a stand-by competent leader available if accident befell the president.[30]

Thus, the 25th Constitutional Amendment was adopted in 1967, to fill the vice presidential office. As it turned out, it was quite a smart move considering what was about to happen in the 70s.

In October 1973, Vice President Spiro T. Agnew, facing accusations of bribery and tax evasion during his tenure as a governor of Maryland (1967–1969), was forced to resign. Section 2 of the 25th Amendment, stating that "whenever there is a vacancy in the office of the Vice-President, the President shall nominate a Vice-President who shall take office upon confirmation by a majority vote of both Houses of Congress," had to be applied. Gerald Ford was nominated in October 1973, to be confirmed by the Senate in November, and by the House of Representatives in December. The provisions of the 25th Amendment were implemented, and Ford became the first appointed American Vice President. Only after another nine months, in August 1974, following Richard Nixon's resignation due to the Watergate scandal, Ford assumed

[28] For a transcript of the Carter-Ford debate, see CPD, 1976 Debate Transcript, http://www.debates.org/pages/trans76b.html (accessed March 29, 2009).
[29] Davis, *American Presidency*, 364.
[30] Cronin and Genovese, *Paradoxes of the American Presidency*, 315.

the highest U.S. Office. He became the only President not to come from a national election. Moreover, to implement the mentioned Amendment, another "unselected" candidate, former New York governor Nelson Rockefeller, became Vice President.

Thus, in the 1976 presidential campaign, it was not only important who was the standard bearer, but the eyes of many voters also turned to the vice presidential candidate. Debates would be an excellent opportunity to learn more about the candidates, take a closer look at their leadership skills, political views and life experience.

There are several purposes of Veep debates. Vice presidential debate is a stage where:

> a candidate must demonstrate fitness to assume the presidency—an office he or she is not actually seeking; define his or her role as a vice president and demonstrate fitness for that office as well; contribute to a better understanding of the presidential running male's virtues; respond to attacks on policies and character in a manner that the presidential nominee could not; and lodge more direct attacks on the opponents, especially the presidential running mate, than could be made in a direct matchup between presidential nominees.[31]

As can be seen, these purposes are very similar to the running mates' general campaign duties. However, since not much campaign time is devoted to exposing the vice presidential candidates, the debate is the best moment to defend the standard bearer, attack the opponent and demonstrate leadership qualities in front of the nationwide audience.

The first two vice presidential debates, of 1976 and 1984, were sponsored by the League of Women Voters. Since 1988 the debates have been sponsored exclusively by the Commission on Presidential Debates (CPD), a nonprofit, nonpartisan corporation. It was established in 1987 "to ensure that debates, as a permanent part of every general election, provide the best possible information to viewers and listeners. Its primary purpose is to sponsor and produce debates for the United States presidential and vice presidential candidates and to undertake research and educational activities relating to the debates."[32] The CPD is also responsible for the format of the debates. In the debates of 1976, 1984 and 1988, running mates were asked questions by a group of panelists. In 1992, both in

---

[31] Carlin and Bicak, "Toward a Theory"; see also "The Vice Presidential Debates of 1976, 1984, and 1988," in Hinck, *Enacting the Presidency*, 167–212.

[32] CPD, *Our Mission*, http://www.debates.org/pages/about.html (accessed March 29, 2009).

presidential and vice presidential debates the single moderator format was introduced, and is now standard practice.

The very point of Veep debates was articulated during the 1988 clash, when Dan Quayle and Lloyd Bentsen were running mates. In an opening statement Bentsen remarked that "The debate tonight is a debate about presidency itself, and a presidential decision that has to be made by you [the voter]. The stakes could not be higher."[33]Although Bentsen used it to focus on Quayle's presidential qualifications, or, rather their lack, his remark can be viewed as a motto concerning the importance of first selecting, and then electing, the vice president. It was also this 1988 debate that produced one of the most famous moments in the history of vice presidential debates, when Dan Quayle, trying to defend his inexperience, compared himself to John Kennedy, saying, "It is not just age; it's accomplishments, it's experience. I have far more experience than many others that sought the office of vice president of this country. I have as much experience in the Congress as Jack Kennedy did when he sought the presidency." Bentsen responded quickly and passionately, "Senator, I served with Jack Kennedy, I knew Jack Kennedy, Jack Kennedy was a friend of mine. Senator, you are no Jack Kennedy."[34] What was even more awkward, Quayle was unable to answer the simple but crucial question concerning the office he was running for. Asked what steps he would take if he suddenly had to assume Presidency, instead of responding briefly, Quayle started deliberating: "First, I'd say a prayer for myself and for the country that I'm about to lead. And then I would assemble his people and talk. And I think this question keeps going back to the qualifications and what kind of Vice President in this hypothetical situation, if I had to assume the responsibilities of President, what I would be."[35] Even if it is hypothetical, it is still one of the basic and key questions which the second office candidate should be able to answer. Not only should he have been prepared for a straight answer, but he should also have had a detailed plan of action in case such a hypothetical situation occurred in real life. Quayle's performance cost the GOP ticket, led by George Bush senior, several percentage points of support.[36]

The biggest blunder, however, or at least one that certainly had some influence on the election, came in the first, 1976 vice presidential debate. Republican Bob Dole called the wars in which the United States had

---

[33] For a transcript of the Quayle-Bentsen debate, see CPD, 1988 Debate Transcript, http://www.debates.org/pages/trans88c.html (accessed March 29, 2009).

[34] Ibid.

[35] Ibid.

[36] Hines, *Gore-Kemp*, 1.

participated in the 20th century—WWI and WWII, the Korean War, and the Vietnam War—"Democratic Wars," as the decisions to participate were made by Democratic administrations. This must have had an effect on viewers, around 43.2 million that night.[37] After losing the elections, Gerald Ford named dropping Nelson Rockefeller and selecting Dole as his running mate "the biggest mistake of his life." It cost him not only crucial electoral votes in the state of New York, as it turned out, but also the choice made by undecided voters from all around the country. The debate revealed the dark side of Robert Dole, which resulted in 5 percent switching to Carter because they did not like Dole.[38] As for Dole, it took him years to live down the hatchet-man reputation he had gained.[39]

Another interesting event was the 1984 debate between George Bush Senior and Geraldine Ferraro. It was important for both candidates. For Bush, probably the best qualified vice president in history at that time, it opened the door for the 1988 Republican Party nomination. Ferraro, on the other hand, as the first female vice presidential candidate in history, was under closer scrutiny, the objective of which was to see how a woman would perform in the stressful situation of a public debate of such caliber. Because the vice presidential debate usually occurs after the first and before the second and possibly third presidential debate, the running mate can also strengthen and refresh the position of the ticket. It was Bush's main task in the debate—Ronald Reagan had performed poorly in his first debate with Walter Mondale, and Bush was to restore Reagan's image as an effective leader.[40] Before the debate, *The New York Times* pointed out that "the ultimate reaction to the Vice-Presidential debate may play a substantial role in determining the pace and tone of the Presidential contest."[41] Defending and praising Reagan, as well as skillfully diverting attention from the issue of Reagan's age, Bush was able to restore citizens' confidence in the president's leadership.[42] Ferraro, although she did not dominate the debate, performed well. Still, she was unable to defend the Democratic ticket from landslide defeat, as the Republicans won in every state, except the District of Columbia and Mondale's home state of Minnesota. With his performance in the debate, Bush strengthened his self-esteem and earned new supporters within the GOP, which secured his

---

[37] CPD, 1976 Debates, http://www.debates.org/pages/his_1976.html (accessed March 29, 2009).
[38] Goldstein, *Modern American Vice Presidency*, 132.
[39] Hines, *Gore-Kemp.*
[40] Hinck, *Enacting the Presidency*, 180.
[41] Gailey, "Bush-Ferraro Debate," A1.
[42] Hinck, *Enacting the Presidency*, 180.

nomination in the 1988 election. Nevertheless, as communication theorist Judith Trent suggests, the spotlight on the debate had less to do with what they said and did than it had with the fact of their gender.[43]

Also the 1996 Veep debate, between Al Gore and Jack Kemp, considered the best vice presidential debate to date, opened new perspectives for the participants. They had such an impressive, issue-oriented and interesting discussion that some even speculated that instead of being a mere vice presidential debate, it might turn out to have been nothing less than a rehearsal of their future presidential clash in the election of 2000.[44]

It is worth taking a closer look at the fact that usually it is very hard to point to the clear winner of vice presidential debates—significantly, it is the media that will tell us who the winner is. Watching the debate on our own, it is quite difficult to judge who performed better, until we read or listen to the comments, mainly provided by the media. More than anything, running mates are expected not to harm the ticket by attacking the opponents too vehemently (as did Dole in 1976), as it might be used against them. The point is less to win the debate than to expose the opponent to see if perhaps he performs poorly. On the other hand, sometimes those who were not necessarily brilliant in actual debate become winners. Republicans Quayle and Dick Cheney got more out of the debates of 1992 and 2000, consecutively, because they performed better than they had been expected to. Cheney, for instance, managed to improve his dark image of former White House Chief of Staff, Secretary of Defense and Halliburton CEO. Before the debate, only 40 percent of likely voters had a favorable impression of him.[45] But in the debate he was a "nice guy." According to someone who helped Joseph Lieberman, running on a Democratic ticket with Al Gore in 2000, the senator assumed that Cheney would play the traditional role of the vice presidential candidate, i.e. attack. When Cheney was a model of civility, Lieberman was left unprepared to score his points and lost much of the edge.[46] On the other hand, however, Cheney's performance in 2004 gave boost to John Edwards in much the same way—people had expected much more from an incumbent vice president and one of the architects of Bush's policy.

The issues discussed in such debates refer mostly to contemporary politics. Candidates try to present their programs and praise the standard bearers. Whenever there is an incumbent ticket, the sitting vice president

---

[43] Trent, *1984 Bush-Ferraro*, 121–44.
[44] Shales, *Al Gore's Gentle Breeze*, D01.
[45] Gledhill, "State Survey," A5.
[46] Allen and Harris, *Debate Assumes New Importance*, A01.

will always defend his administration record and state what they want to propose and accomplish during the second term, while the opponent will try to downplay the achievements of the previous four years. If the ticket is a challenger, their rivals will focus on political experience, the voting record, the candidates' opinions on different issues from long ago, in the primary campaign and in the campaign after securing their nominations.

To conclude, it should be emphasized that the effect of the vice presidential debates is rather small. In a close election, like in 1976, the debates can play an important role. Otherwise, however, they matter only to a minority of voters in helping them to decide which ticket to vote for.

For instance, on the morning after the 1984 debate, Bush delivered a speech in the port in New Jersey. As he left the docks, he whispered to one of the union officials, "We tried to kick a little ass last night."[47] Only after saying it did he realize that his microphone was still on and it had picked up his words. He later defended his remark, describing it as "an old Texas football expression."[48] Reagan and Bush won in a landslide.

The Bush-Quayle ticket won in 1988 despite Quayle's inability to overcome all doubts about his readiness for either the presidency or vice presidency.

Although from 1976 to 2004 seven vice presidential debates were watched by the combined number of about 300 million Americans, they seem not to pay much attention to who is elected vice president. This approach to vice presidency is rather inappropriate. Imagine that the President of the United States dies the day after the inauguration. Of course, it is only hypothetical and highly improbable, but not impossible. It could turn out that for the next four years the country would be run by a person whom the U.S. citizens, as well as the world leaders and public opinion, hardly know. Among the first things which they all should probably do to know him better is watching the vice presidential debate from the last election.

Works Cited

Allen, Mike, and John F. Harris. "Debate Assumes New Importance." *Washington Post*, October 4, 2004.

Ambrose, Stephen. *Nixon: Education of a Politician, 1913–1962*. New York: Simon and Schuster, 1987.

---

[47] Trent, *1984 Bush-Ferraro*, 133.

[48] Ibid., 134.

Carlin, Diana B., and Peter J. Bicak. "Toward a Theory of Vice Presidential Debate Purposes: An Analysis of the 1992 Vice Presidential Debate." *Argumentation and Advocacy* 30, no. 2 (Fall 1993): 119–30. http://ebscohost.com.

Cronin, Thomas, and Michael Genovese. *The Paradoxes of the American Presidency*. New York: Oxford University Press, 1998.

Davis, J. W. *The American Presidency*. Westport, CT: Praeger, 1995.

Diggins, John P. *John Adams*. New York: Times Books, 2003.

Dover, E. D. *Presidential Elections in the Television Age, 1960–1992*. Westport, CT: Praeger, 1994.

Friedenberg, Robert V., ed. *Rhetorical Studies of National Political Debates, 1960–1992*. Westport, CT: Praeger, 1994.

Gailey, Phil. "Bush-Ferraro Debate Today Seen as Altering Campaign Dynamics." *New York Times*, October 11, 1984.

Gledhill, Lynda. "State Survey Finds Cheney Got Boost from Debate." *San Francisco Chronicle*, October 11, 2000.

Goldstein, Joel. *The Modern American Vice Presidency*. Princeton, NJ: Princeton University Press, 1982.

Graber, Dorris. *Mass Media and American Politics*. Washington, DC: Congressional Quarterly Press, 1980.

Hatfield, Mark O. *Vice Presidents of the United States, 1789–1993*. Washington, DC: U.S. Government Printing Office, 1997. http://www.senate.gov/artandhistory/history/resources/pdf/theodore_roosevelt.pdf (accessed March 24, 2009).

Hinck, Edward. *Enacting the Presidency: Political Argument, Presidential Debates, and Presidential Character*. Westport, CT: Praeger, 1993.

Hines, Cragg. "Gore-Kemp Face-off to Open Door on 2000." *Houston Chronicle*, October 9, 1996.

Humphrey, Hubert H. "Changes in the Vice Presidency." *Current History* 67 (1974): 58.

Light, Paul. *Vice-Presidential Power: Advice and Influence in the White House*. Baltimore: Johns Hopkins University Press, 1984.

Shales, Tom. "Al Gore's Gentle Breeze." *Washington Post*, October 10, 1996.

Smith, Hendrick. *The Power Game: How Washington Works*. New York: Random House, 1988.

Trent, Judith S. "The 1984 Bush-Ferraro Vice Presidential Debate." In Friedenberg, *Rhetorical Studies*, 121–44.

Windt, Theodore O., Jr. "The 1960 Kennedy-Nixon Presidential Debates." In Friedenberg, *Rhetorical Studies*, 1–28.

Young, Donald. *American Roulette: The History and Dilemma of the Vice Presidency*. New York: Holt, Rinehart, and Winston, 1972.

Still Cheating Us Softly: Video News Releases (VNRs) Masquerading as Solid News in American Media[1]

Małgorzata Gajda-Łaszewska

Invisible, yet influential, set at the crossroads of information and promotion, news releases are a standard tool of PR industry. Their task is to reach the media, to get their attention, and to create publicity for a product, event or initiative. Yet, the combination of new technologies, clutter in the world of advertising and the condition of the American news media, (the local ones in particular), has created a situation in which news releases have acquired a new role and meaning. Rather than wait for media attention, PR practitioners aggressively pursue their promotional goals and take over the media function of news-making. They become hosts rather than guests on air, and arrange for the media to play a role in their own script of disseminating information about a client. The following discussion is going to focus on the phenomenon of the electronic version of news releases—Video News Releases (VNR)—to trace their recent successful inroads into the news media and the influence of this "fifth column" on the media landscape and their audiences.

The simplest description of Video News Releases (VNRs) would call them "electronic press kits, the video equivalent of the written press release,"[2] aiming at providing journalists with story ideas, pictures, quotations and background material for developing independent reports. Kara Rowlands specifies them further, stating that they "are television packages resembling newscasts—usually produced by large corporations, organizations or even the government—that are sent out to TV stations

[1] Extensive resources on VNRs (actual corporate videos and their appropriations by American newscasters) can by found at the Web site of the Center for Media and Democracy, http://www.prwatch.org/.

[2] Jamieson and Campbell, *Interplay of Influence*, 142.

with the hope that reporters will incorporate all or some of the footage into a news segment."[3] Obviously they are sent free of charge for unrestricted use. Thus, "they provide a timely and inexpensive alternative for news directors in rounding out their newscasts, which are costly to produce."[4] Typically, VNRs come packaged with two versions of the story which the PR firm is trying to promote. The first version is fully edited, with voiceovers already included or with a script indicating where the station's local news anchor should read his or her lines. The second version is known as "B-roll," and consists of the raw footage that was used to produce the fully-edited version. The receiving station can edit the B-roll footage itself, adding own script and narration, or combine it with footage received from other sources. "B-roll . . . gives journalists full control to utilize the video and create their own narrative. From a practical standpoint, using B-roll provides significant benefits to news producers and the PR industry. Stock footage has a longer shelf life and is less costly to produce than voiced-over reports, thus providing a greater return on investment. The media also has greater receptivity to B-roll content."[5]

As far as VNR history goes, they were "inefficient as long as they relied on videocassettes as their means of transport. Cassettes began to clutter newsrooms, unwatched and unaired. In the mid 1980s, public relations firms began switching to satellite transmission of news feeds."[6] Then the situation changed for a couple of reasons. Firstly, PR department-media communication became more efficient and the ties between the two seemed to strengthen; so much so that the dissemination of VNRs increased dramatically. There is no exact data as to the scope of VNRs produced yearly but the "estimates place the number of VNRs distributed each year at between 5,000 and 15,000"[7] (a typical newsroom may have 10-15 VNRs available a day[8]). This brought into being an entire VNR industry with two companies—D S Simon and Medialink—leading the way. Secondly, PR practitioners found a vulnerable spot in broadcast organizations and momentarily caught the opportunity to use it to their own advantage. Paradoxically, the advent of cable, which resulted in the multiplication of TV channels, fostered, rather than prevented, further decrease in the length of a sound bite. Those who believed that cable

---

[3] Rowland, "FCC Fines Use of VNRs."

[4] McCleneghan, "Are VNRs All They're Cracked Up to Be?"

[5] Purushothaman, "Truth about VNRs."

[6] Jamieson, *Interplay of Influence*, 141.

[7] Ibid.

[8] A study by Mark Harmon and Candace White at the University of Tennessee, quoted in Farsetta and Price, "Fake TV News: Introduction."

would usher the era of profound, comprehensive and detailed reports increasing the viewers understanding of issues were gravely disappointed. A long-time CNN medical correspondent complained:

> Many of us believed the growth of television channels would mean more time to tell stories with context. Time to explain caveats and side effects of new medical treatments, not just the wonders promised in press release headlines [*sic*]. Time to spell out specific strengths and limitations of health studies, and to avoid whipsawing viewers with research snippets that seem to contradict each other, simply because key details were left on the cutting-room floor. We were right . . . for a while. Twenty years ago, the typical length of a CNN medical story was almost 3 minutes, and even longer in the weekend wrap-up program. But as cable penetration increased and channels proliferated, a strange thing happened: Time contracted. Between CNN and Headline News, we were transmitting 48 hours of news a day, 14 days a week, but we were told to trim story length to 2:30 . . . then 2:15 . . . then 2 minutes. As we spiraled down I bailed out. Cable surrendered to TV's tendency to churn out quick snacks rather than balanced meals . . . This blizzard of medical voiceovers doesn't inform; it dazes.[9]

With further decrease of news item size (health stories today are usually shorter than 33 seconds[10]), stations found themselves in the situation in which they had more time to fill and had to compete fiercely for viewers. Moreover, the trend plaguing the network news and bearing on the news quality is that whereas the number of reporters is decreasing, they are supposed to file more reports. As a result, "the workload per reporter is increasing."[11] Diane Farsetta from the Center for Media and Democracy quotes the findings of TV newscasts monitoring which show that, on average, in 1996, fifty-one reporters filed thirty-five stories, while in 2005, forty four reporters prepared thirty-nine stories.[12] For local news providers, bound by shoestring budgets, plagued by cutting reporters' opportunities to pursue ambitious stories, and generally decreasing the number of journalists in the newsroom, the task to keep the offer attractive seemed really challenging. In this situation professional-quality, attractive and easily available VNRs provided a tempting alternative. Thus, it is generally assumed that the saturation of the media with VNRs has mostly affected the local news market. This, however, was not confirmed by the

---

[9] Holtz, "Targeted Treatment."

[10] Ibid.

[11] Farsetta and Price, "Fake News: Introduction."

[12] Ibid.

Center for Media and Democracy survey, carried out twice in 2006, which seemed to suggest that all markets actually employed PR materials, corporate or other: "The list of worst offenders includes Clear Channel, New Corp./Fox Television, Viacom/CBS, Tribune Co. and Sinclair Broadcast Group—whose Oklahoma City affiliate was caught airing VNRs on six separate occasions."[13] The research group tried to show the relation between large conglomerate ownership of the stations and VNR usage. Free Press campaign director Timothy Karr claims that "the evidence suggests a strong connection between media consolidation and the broadcast of deceptive, pre-packaged propaganda . . . When all station owners care about is profit margins, fake news can prove irresistible. After all, VNRs are free. Reporting news that's meaningful to local communities isn't."[14]

At this point a question arises: if VNRs are just an electronic version of the well-known press release, why do they stir so much emotion and raise controversy? Over the last two and a half decades they have been a bone of contention between media watchdog groups, such as Free Press or the Center for Media and Democracy, PR firms and the media outlets which they address.

The most problematic issue concerns the media's non-disclosure of VNRs' origin. Sponsored clips frequently find their place on the news, yet they masquerade there as regular reports prepared by the staff journalists rather than promotional material put together by PR specialists, according to their script and paid for by the corporate (usually) sponsors. It is vital to add that if not identified by the anchor, VNRs are practically indistinguishable from the regular news reporting for an average viewer. PR companies, although aggressively pursuing their contacts with media people and making inroads into the newsroom, wash their hands when it comes to this accusation, claiming that their materials usually contain clear disclosure throughout or at least at the end of the clip, but it is frequently removed by the reporter while the footage is utilized as the station's own. Not that PR practitioners are unhappy with such a state of affairs; they just do not take responsibility for it. Moreover, they blatantly claim that, firstly, news media with their dire financial situation are doomed to such an arrangement and secondly, that it is not hazardous to viewers but serves as a source of valuable and unique information as well as footage of which the audiences would otherwise be deprived. Besides, the reasoning goes,

[13] Farsetta and Price, "Fake TV News: News Release."

[14] Quoted in Farsetta and Price, "Fake TV News: News Release."

the productions do not contain any false information. Thus, they do not really deserve being dubbed "fake news." Kevin E. Foley blatantly retorts:

> CMD [Center for Media and Democracy] would have us believe that some great social harm is being done if a VNR isn't attributed, but if the newscaster airs a story that holds the viewer's attention and the viewer walks away informed or entertained, who has been hurt? Newscasters decide the editorial value of the content we offer as they've always done, so it's not as though there aren't safeguards in place. Indeed, one of the common practices in print journalism is to lift quotes and even blocks of copy from press releases. Reporters from local weekly newspapers to *The New York Times* do it every day without ever disclosing to the reader the source of the material. And you know what? Nobody cares.[15]

As shocking as the admission may seem, the fact that everybody does that does not necessarily make the practice more acceptable or ethical.

VNRs have their roots in PR people's search for paths of corporate promotion other than advertising in the media world. A popular belief has it that in the contemporary advertising clutter it is getting harder to get through with one's message. Moreover, due to constant exposure, audiences have developed some immunity mechanisms and do not take ads at their face value any longer. Viewers have become much more suspicious and cynical toward the corporate sponsored messages. On the other hand, this form of publicity lost its "value for money" factor with the prices of TV commercials skyrocketing.

All this led marketing specialists to look for alternatives, to find such loopholes where companies could garner advantageous visibility. The most obvious tactic was to try to make ads look unlike ads. But what? The news programs tended to enjoy comparatively better receptivity and respect. If we add to this the fact that most adult Americans still get their information about the world primarily from television,[16] we should not be surprised to see the attempts to link or mesh the two formats. Research has been carried out; seemingly, the most promising format for disseminating corporate messages were the newscasts. They were utilized by PR firms to the advantage of their clients in two basic ways. One was making original reports paid for by the client or preparing corporate videos and then buying spots on networks to air them as infomercials, which viewers easily

---

[15] Foley, "Are Video News Releases."

[16] The Harris Poll of February 2006 reveals that "three-quarters (77%) of adults watch local broadcast news, and 71 percent say they watch network broadcast or cable news several times a week or daily." Harris Interactive, "Seven in 10 U.S. Adults."

confuse with news updates. Thus, ads were made to look like news. The tactic has paid off, to some extent. It has even acquired its own name: "branded journalism."[17] The other trend went even further. Here, PR professionals were supposed to plant the same corporate messages on the real news, this time for free. This option offered the desired exposure to the issue/product or a company, simultaneously making the authority of the newscast "bleed" to the material; besides, it was cheaper to create and place. Since not only corporations embrace this tactic, Scott Ferguson, chief of The Army Reserve Public Affairs Broadcast Operations, explains its advantages:

> All of this media coverage, whether local or national, has a large payoff . . . "If a 30-second commercial on Fox News in the morning costs $15,000 and you have two Army Reserve soldiers . . . on Fox for five minutes, you're getting $150,000 worth of airtime . . . Let's say a television station in Indianapolis charges $1,000 to $1,500 for a 30-second commercial during their 6 o'clock news. Say they run a 1:30 package [using Broadcast Operation footage] at 6 p.m., do a sidebar at 10 p.m., and maybe they run it again at 5 a.m. . . . That is $6,000 of positive airtime from one station, let alone the other three in the market."[18]

What should be added is that the cost of producing such a film is also smaller than that of creating a commercial. Medialink Chairman Laurence Moskowitz boasts: "We can produce a 90-second newscast for the cost of catering a traditional 30-second-spot—and we can turn it around for hours."[19] Moreover, some VNRs resulting in the incorporation of marketing footage in the newscasts allow numerous corporations to bypass restrictive regulations or outright bans binding them in case of pure advertising, e.g. talking about hard liquor in the broadcast media or omitting side effects of promoted drugs. All in all, once PR firms realized that the media people are eager to use them, the entire VNR industry began to thrive.

A corporate PR department or a contracted outside PR firm prepares a film covering an issue of vital importance to the company, or presenting a new product or service offered to the market. As typical PR material, the tone of the video always praises the sponsor and underscores its contribution to the community or society at large. Display of the company

---

[17] Mandese, "Art of Manufactured News."

[18] Scott Ferguson, chief of The Army Reserve Public Affairs Broadcast Operations in DiBenedetto-Kenyon, "Video on Reserve." For more examples, see Simon, "VNR/SMT Packages."

[19] Mandese, "The Art of Manufactured News."

logo or the name of the product constitutes a significant portion of the material. The narrator provides the story which usually ends with a disclosure of the sponsor. Such film is frequently accompanied by a B-roll. "Broadcast PR firms announce their VNR offerings to TV news producers through phone, fax [in his 1996 market offer, D S Simon promised fax alerts to more than 750 reporters, by name and beat assignment utilizing our customized data base'[20]] and e-mail pitches."[21] Then the package of the two videos is forwarded to the media to get the attention of reporters of a particular beat. Thus, the reception is quite specifically targeted. If we add to this mixing of PR people and reporters during various events and occasions, we can infer that many of them know personally those whom they "feed" their material. Another, more passive way of distributing VNRs is putting them into a news source available for reporters, again organized by beats or topics, from which they frequently draw their ideas (as we can see, often it does not stop at that). These bases are maintained by reliable news institutions such as AP or CNN or separate commercial entities such as Pathfire or Digital Media Gateway. Supposedly, it is not infrequent for reporters themselves to turn to PR companies and ask them for footage. This is, however, extremely hard to prove since "there are truly hundreds of examples of self-denial on the part of broadcasters when it comes to admitting that VNRs are used"[22] at all.

Once a reporter gets a VNR he or she has a few options at hand: 1) to use it as a story idea, a starting point to conduct one's own, independent investigation. Although the most ethical and honest, it is not the option most often adopted, due to financial and temporal constraints which a typical newsroom faces; 2) to use it as it appears in the news report, disclosing the sponsor. Yet this, as we know, also happens very rarely because it is not appealing to either side (sending or receiving), the viewer excluded. Viewers' interests, however, seem not to have been the main issue in this case, anyhow; 3) to use it as a part of the station's own news report without disclosing the sponsor or the producer. The main trap is that actually

> neither the PR industry nor the TV news media have any incentive to discuss this topic in places where the public will notice. Neither wants the public to know the truth about where their "news" is coming from. For the PR firms, full disclosure weakens the effectiveness of their propaganda.

---

[20] Simon, "VNR/SMT Packages."

[21] Farsetta and Price, "Fake TV News: Frequently Asked Questions."

[22] George Glazer, a senior vice president of Hill and Knowlton quoted in Rampton, "Fake News? We Told You So."

> For the TV stations, admitting that they use VNRs in place of actual reporting would be the equivalent of a bakery admitting that it uses sawdust in place of wheat in its bread: cheap, non-nutritious filler that serves only to add worthless bulk to the product.[23]

This, however, is not the end of the story. Stations not only fail to disclose the corporate source of their news segments but also try to disguise them further as "real news," not fake ones.

Daniel Price from the Center for Media and Democracy has identified main techniques employed in the process. They consist of "adding network-branded graphics and text overlays" (done in all clips under CMD scrutiny), revoicing the material (in more than 60 percent of cases) and replacing "the original VNR narration with their own," frequently using exactly the same script, or introducing a publicist from the clip as if he or she were a station's reporter (in almost half of the VNR broadcasts surveyed).[24] Such tools further obscure the borders between the news report and the VNR and render the two almost indistinguishable. And, despite some claims that nobody really cares if a viewer is informed or at least entertained, the mystification is of major importance when we realize that the source is not a disinterested party. Jonathan S. Adelstein, FCC Commissioner, commented: "The public is misled by individuals who present themselves to be independent, unbiased experts, but are actually shills promoting a prepackaged corporate agenda."[25] This is exactly why VNRs pose a hazard to the integrity of the media as news-gathering institutions in a democratic society. Here, we also come to see that a semblance of VNRs to their older sisters, press releases, is only apparent. The main difference seems to be that while press releases were "a tool to attract the attention of journalists, VNRs are often used to replace journalists entirely."[26] The Center for Media and Democracy Study shows that only in 7 out of 87 cases "stations added independently-gathered footage or information to the segment," while in 31 instances there was not a single edit of the material.[27] This is as much as to say that the media assume and accept the corporate point of view. Reporters do not try to provide any verification, check the facts, balance the message with independent footage, let alone question the issue altogether. A serious effect of VNRs' weaving into the newscasts is "the exclusion of important,

---

[23] Rampton, "Fake News? We Told You So."
[24] Farsetta and Price, "Fake TV News: Frequently Asked Questions."
[25] Pescatore, "FCC Inquiry."
[26] Farsetta and Price, "Fake TV News: Frequently Asked Questions."
[27] Ibid.

relevant information that challenges or weakens client's messages."[28] Since corporate VNRs save TV producers' lives in many cases they are not going to bite the hand that rescued them. What it comes down to is a neat symbiosis between the media and the corporate sector (as most VNRs derive from corporations). This results in the slow disappearance of negative stories which reveal various types of corruption and relegating investigative reporting, the most noble genre of journalistic practice, to the fringes of the media universe.[29] In the new economy, media stoop to adopting a PR perspective, with corporations setting the agenda of what is going to fill the newscasts. Alarmingly, we are witnessing a clear role reversal: corporations write scripts for the media to follow. Furthermore, we have to realize that in fact PR productions rarely, if at all, have real news value and they take up the time that should be devoted to much more valid, yet frequently controversial issues influencing the community. What we do receive, however, is an ideal, sugar-coated world created out of corporate interest. "More and more frequently, Americans can't tell PR and press apart . . . Increasingly, the press and public relations practitioners have begun to take on interchangeable roles. This is not only unethical, it denigrates both professions."[30]

Finally, the question arises: why has the issue of a widespread usage of VNRs remained invisible for such a long time? The president of the Radio-Television News Directors Association (RTNDA), Barbara Cochran, expressed her somewhat cynical opinion on VNRs, stating that they are "kind of like the Loch Ness Monster. Everyone talks about it, but not many people have actually seen it."[31] Her answer is simple and cruel at the same time. It is not in the interest of corporations, PR firms or the media to disclose the shameful procedure, so the public is mostly unaware of the cheating going on.

Two media watchdog groups, the frequently quoted Center for Media and Democracy and Free Press have played a vital part in publicizing the issue. They organized the "Take Action Now!" campaign, where they gathered signatures of 40,000 Americans interested in lobbying the law making bodies (FCC) to take a stand and establish clear standards for VNR disclosure. They have campaigned, for many years, demanding clear labeling of all VNRs on screen so that viewers can clearly know the difference. One of the forms of their actions were formal complaints filed

---

[28] Farsetta and Price, "Still Not the News."

[29] For a very good example of TV-hospital partnerships, see Lieberman, "Epidemic."

[30] Lordan, "Defining Public Relations," 41.

[31] Farsetta and Price, "Fake TV News: Widespread and Undisclosed."

with the FCC, requesting an investigation of the procedure. What public interest groups want to achieve is that "all VNRs be accompanied by a continuous, frame-by-frame visual notifications and verbal announcements disclosing their resources. They also recommended broadcasters be required to file monthly public reports detailing their use of government or corporate-sponsored material."[32]

As a response to the Center for Media and Democracy and Free Press activities, FCC issued a public notice in April 2005 calling for full disclosure of VNRs on the news, referring to "the principle that listeners and viewers are entitled to know who seeks to persuade them with the programming offered over broadcast stations and cable systems."[33] The notice clearly stated that "whenever broadcast stations and cable operators air VNRs, licensees and operators generally must clearly disclose to members of their audiences the nature, source and sponsorship of the material that they are viewing."[34] This FCC call was supported by the Radio and Television News Directors Association suggesting that voluntary disclosure is the best media policy with respect to VNRs. Yet, a survey published by *PR News* in January 2006 showed that PR professionals "largely ignored calls for full disclosure."[35] The majority of polled PR specialists (53%) did not change their disclosure policies to make their communications with the media more open when using a spokesperson or producing a VNR. Moreover, 46% of respondents "planned to repeat the same level of VNR usage for 2006 as they did for 2005."[36] Mark Hirsch says that "for every PR professional that provides the media with full disclosure, there are three failing to meet the RTNDA's requirements."[37] The PR executives and broadcast associations tried to downplay the significance and gravity of the procedure. Obviously, they have strongly opposed meaningful disclosure policies, claiming that voluntary codes of conduct already binding them are sufficient, especially as VNRs are said to be used only as a supplement to original newsroom reports and on rare occasions. Besides, PR people argued that, in many cases, VNRs provide footage which, while valuable for the audiences, is otherwise unavailable to the newsroom. The most solid argument threatened that the enforcement of any stricter disclosure

---

[32] Farsetta and Price, "Fake TV News: News Release."
[33] Federal Communications Commission, *FCC 05-84*.
[34] Ibid.
[35] *PR Newswire*, "Most PR Pros," 1.
[36] Data from *PR Newswire*, "Most PR Pros."
[37] Ibid.

policies would infringe editorial independence of the newsroom and would imply an attempt at governmental censorship.[38]

Nevertheless, the FCC warning that "We will take appropriate enforcement action against entities that do not comply with these rules"[39] did not remain an empty threat. It materialized in September 2007 in a groundbreaking decision to fine Comcast Corporation for five identified instances of undisclosed usage of VNRs on the news program. The Commission's Enforcement Bureau proposed a $4,000 fine for Comcast regional cable net CN8 for each recognized footage of unidentified VNR ($20,000 total). All five fined clips were aired on the *Art Fennell Reports* program.[40] CN8 retorted that "the VNR rule applies only to broadcasting—but even if the rule is valid, no sponsorship ID is necessary as CN8 and its employees didn't receive any consideration in exchange for airing the material,"[41] while the FCC rule covered in the Telecommunication Act of 1934 clearly mentions the cases when "money, service, or other valuable consideration is directly or indirectly paid, or promised to or charged or accepted by, the station so broadcasting."[42] But the director of George Washington University's journalism program aptly pointed out that "because of the high cost of compiling video for a newscast stations that accept outside video are in effect accepting in-kind contribution from that source."[43]

Among the media and in the PR sector the reaction to the decision was instant and very fierce. Obviously, they defended their practices and denounced the ruling. A blogger very succinctly summed up the attitude as "wrapping oneself in the flag."[44] In their defense against the FCC step and in their lobbying, the main opposition leader, RTNDA, joining forces with a hastily formed "coalition" of broadcasters (many of whom are under investigation from the FCC), referred to the First Amendment protections. The RTNDA news release reads:

---

[38] Farsetta and Price, "Fake TV News: Recommendations."

[39] Federal Communications Commission, *FCC 05-84*.

[40] The clips included Nelson's Rescue Sleep; General Mills' "Wheaties Fit to Win the Challenge"; Allstate Insurance's life insurance programs; Trend Micro Software's "Remote File Lock"; and General Mills' "Bisquick 75th Anniversary" in Hearn, "FCC Hits CN8."

[41] *Cablefax Daily*, "Fake News? FCC Bureau."

[42] Federal Communications Commission, *FCC 05-84*, 2.

[43] Farsetta and Price, "Fake TV News: Recommendations."

[44] Media Citizen, "Fake News Flack Ducks"; about D S Simon reaction to the FCC Notice for VNR disclosure.

> The FCC Enforcement Bureau injects the government agency into protected newsgathering and editorial activities to an unprecedented and inappropriate degree, in clear conflict with journalist's First Amendment rights . . . The FCC has embarked upon an extraordinarily dangerous slippery slope toward government censorship, and has imposed a chilling effect on the use of all types of third-party materials that electronic journalists may independently decide are of interest or importance to their viewers.[45]

As a result, RTNDA and a coalition of some 70 state broadcast associations and broadcast group owners asked the FCC to rescind the fine, which, however, did not take place.

Yet, the entire debate and the dynamics of the conflict reveal a certain impertinence of the PR-corporate-media alliance, and a shocking finality, as well as unavoidability of their practices. We witness a strange repositioning and reevaluation of concepts when protection of the audiences and appealing to media ethics is presented as more hazardous than ads masquerading as news. This echoes the statement of Arthur Sylvester, a spokesman for the Pentagon under Kennedy, who said at a news briefing in November 1962: "I think the inherent right of the government to lie—to lie to save itself when faced with nuclear disaster—is basic, basic." The media seem to have a similarly basic right to lie, although the disaster is of an economic rather than nuclear nature.

Works Cited

Cablefax Daily. "Fake News? FCC Bureau Proposes Fine for Comcast's CN8." September 25, 2007. http://proquest.umi.com/.

DiBenedetto-Kenyon, Laura. "Video on Reserve." *Government Video*, September 2005. http://www.governmentvideo.com/article/30026 (accessed March 20, 2009).

Farsetta, Diane, and Daniel Price. "Fake TV News: Widespread and Undisclosed." Center for Media and Democracy. *PR Watch* 13, no.2 (April 6, 2006). http://www.prwatch.org/fakenews/execsummary (accessed March 20, 2009).

—. "Still Not the News: Stations Overwhelmingly Fail to Disclose VNRs." Center for Media and Democracy. November 14, 2006. http://prwatch.org/fakenews2/execsummary (accessed March 20, 2009).

[45] RTNDA, "RTNDA and 70 Media Organizations."

Federal Communications Commission (FCC). *Public Notice FCC 05-84*. Washington, DC, April 13, 2005. http://www.fcc.gov.

Foley, Kevin E. "Are Video News Releases All Bad?" *Broadcasting and Cable*, May 1, 2006. http://proquest.umi.com/.

Harris Interactive. "Seven in 10 U.S. Adults Say They Watch Broadcast News at Least Several Times a Week." The Harris Poll #20, February 24, 2006. http://www.harrisinteractive.com/harris_poll/index.asp?PID=644.

Hearn, Ted. "FCC Hits CN8 Over Video News Releases." *Multichannel News*, October 1, 2007. http://proquest.umi.com/.

Holtz, Andrew. "Targeted Treatment." *TelevisionWeek*, March 12, 2007. http://proquest.umi.com/.

Jamieson, Kathleen Hall, and Karlyn Kohrs Campbell. *The Interplay of Influence: News, Advertising, Politics, and the Mass Media*. 5th ed. Belmont, CA: Wadsworth / Thomas Learning, 2001.

Karr, Timothy. "Fake News Flack Ducks Behind First Amendment." MediaCitizen, April 15, 2005. http://mediacitizen.blogspot.com.

Lieberman, Trudy. "Epidemic." *Columbia Journalism Review* 45, no. 6 (2007). http://proquest.umi.com/.

Lordan, Edward J. "Defining Public Relations and Press Poles in the Twenty-First Century." *Public Relations Quarterly* 50, no. 2 (2005). http://proquest.umi.com/.

Mandese, Joe. "The Art of Manufactured News." *Broadcasting and Cable*, March 28, 2005. http://proquest.umi.com/.

McCleneghan, J. Sean. "Are VNRs All They're Cracked Up to Be?" *Public Relations Quarterly* 42, no. 4 (Winter 1997–98). http://proquest .umi.com/.

Pescatore, Mark J. "FCC Inquiry Targets Video News Releases." *Government Video*, November 2006. http://proquest.umi.com/.

PR Newswire. "Most PR Pros Still Fail to Disclose." February 21, 2006. http://proquest.umi.com/.

Purushothaman, Shoba. "The Truth about VNRs." *Broadcasting and Cable*, April 18, 2005. http://proquest.umi.com/.

Radio-Television News Directors Association (RTNDA). "RTNDA and 70 Media Organizations Ask FCC to Rescind Fines on Comcast for VNR Use." RTNDA News Release, November 1, 2007. http://www.rtnda.org/pages/posts/rtnda-and-70-media-organizations-ask-fcc-to-rescind-fines-on-comcast-for-vnr-use202.php.

Rampton, Sheldon. "Fake News? We Told You So, Ten Years Ago." Center for Media and Democracy. *PR Watch* 12, no. 2 (March 26, 2005). http://www.prwatch.org/prwissues/2005Q2/toldyouso.html.

Rowland, Kara. "FCC Fines Use of VNRs." *Washington Times*, October 3, 2007. http://www.washingtontimes.com/apps/pbcs.dll.

Simon, Douglas. "VNR/SMT Packages Are a Hot PR Tool." *Public Relations Quarterly* 41, no. 4 (Winter 1996–97). http://proquest.umi .com/.

AMERICAN WAR ON TERROR IN *SLEEPER CELL*: ARAB AND MUSLIM AMERICANS AFTER 9/11

ALEKSANDRA M. RÓŻALSKA

> The Arab world—twenty-two countries, the locus of several world religions, a multitude of ethnic and linguistic groups, and hundreds of years of history—is reduced to a few simplistic images. It is as though American society was to be portrayed solely in terms of cowboys, gangsters and Britney Spears.
> —Marvin Wingfield and Bushra Karaman

> From Zbigniew Brzezinski's vision of the "crescent of crisis" to Bernard Lewis's "return of Islam," the picture drawn is a unanimous one. "Islam" means the end of civilization as "we" know it. Islam is anti-human, antidemocratic, anti-Semitic, anti-rational.
> —Edward W. Said

Introduction

The war on terror, declared by President George W. Bush after the terrorist attacks on 9/11, which have evoked many controversies both in the United States and abroad, has been accompanied by the process of radicalization of already negative attitudes toward Arab and Muslim Americans in American society. These moods have also been reflected in media texts as well as in political language which has been used, also in mass media, to divide society along racial and religious lines into "good American us" and "bad Muslim them." The war on terror demonized the enemy—the terrorists often associated with Arabs and Muslims—that aimed at maintaining a certain level of anger and tension within society. These negative feelings were also strengthened by media images, which, especially in moments of crisis, when society's sense of security is threatened, become important points of reference for public opinion and may arouse fear. However, such images were present in American films and television programs long before the events of 9/11, which is related to a permanent misinterpreting and misunderstanding of Arab and Muslim cultures that are ethnically, politically and religiously different than the dominant American values. Thus, in the first part of this article I briefly refer to the idea of Orientalism, which has influenced the common Western perception of Arab and Muslim cultures. Moreover, it serves as an explanation of their dominant and mostly negative representations in American mainstream culture. Then I analyze the popular American television mini-series *Sleeper Cell* (2005) which, in my opinion, is a successful attempt to challenge the binary oppositions of "good Americans" versus "bad Muslim Arabs" and to a certain extent questions the dominant stereotypical perception of Muslim and Arab cultures.

Orientalism: Looking at the East through Western Eyes

In his famous book published in 1978 Edward W. Said defined Orientalism as a set of false and ethnocentric assumptions which contributed to Western attitudes and perception of the East, Arabs and Islam. These attitudes can be described as permanent, but often subtle and hidden prejudice, which results from the historical relations between the so-called West and the East. Western historians, writers, artists and politicians have underlined the huge differences as well as vast discrepancies in culture and civilization between the two worlds to such an extent that the dialogue and mutual understanding appeared impossible. Additionally, in this opposition between the West and the East it has always been the East which was believed to be weak, chaotic, and in need of constant assistance. With the East having such a position in Western discourse and within the existing power relations, it was easier for Europe and later for the United States to justify colonial ambitions and imperialist expansion. This legitimization of prejudice was achieved also by false—sometimes romanticized and sometimes vilified—images of the East in Western culture. Throughout the book Said underlines that the common image of the Middle East and Asia created by Western scholars, philosophers, or writers for European/American audiences is radically at variance with real cultures of the region: "I believe it needs to be made clear about cultural discourse and exchange within a culture that what is commonly circulated by it is not 'truth' but representations. . . . The value, efficacy, strength, apparent veracity of a written statement about the Orient therefore relies very little, and cannot instrumentally depend, on the Orient as such."[1]

The core of traditional orientalist thinking is a common and deeply ingrained conviction that the people "there" are different from "us" and they do not believe in "our" values.[2] This sharp division between the East and the West underlined by writers, poets, philosophers as well as political theorists served to justify the alleged inferior position of the so-called "Orient mind" in comparison to Western "enlightened" knowledge.[3] Consequently, Orientalism is something more than just a set of false impressions, misleading representations, or "illusionary fantasy"—as Said puts it; it is rather a systematized set of theories and practices maintained and consolidated in Western consciousness across generations.[4]

In this book and his subsequent works Said condemned the West not only for a very poor and simplified understanding of Arab culture and Islam, but also for depriving the East of the possibility to speak with its own voice. He emphasized that political and economic influences of Western countries were not sufficient for them—the East was also symbolically conquered in the sense that the analysis, interpretation and exploration of Eastern cultures, traditions, religion, language, and history, etc. was reserved for Western "reformers." Consequently, the Orient's past was constructed first by Europeans and then by Americans with no respect for its multiple identities and historical complexities. In other words, Euro-American culture became a norm in Western writings whereas the East for centuries was to remain exotic, weak, irrational, less civilized and—later on—deviant. On the other hand, Western civilization has been associated with reason and power, which—in the context of American Manifest Destiny—gives it "moral" right to conquer or expand influences onto the Orient which is—in the Western view—unable to exist without the control and constant guidance from the West.

What, therefore, has been perceived as characteristic for the Orient or Arab cultures? To begin with, they have always been represented as different than the West and, thus, unable to benefit from Euro-American knowledge and to make use of its achievements. Consequently, Arabs were viewed as less intelligent, lacking energy, initiative or willingness for Western-style modernization. The dominant binary oppositions that need

---

[1] Said, *Orientalizm*, 55–56.
[2] Ibid., 14.
[3] Ibid., 31
[4] Ibid., 36.

to be emphasized here are: rational/irrational, civilized/uncivilized, virtuous/depraved, normal/abnormal, childish/adult, culture/nature, rich/poor, developed/backward, etc.[5] However, this is only one aspect of a "true" Arab face as, apart from having difficulties in understanding the Western idea of progress, Arabs are simply malicious people—cunning, lying, suspicious, and vindictive. While Europeans and Americans praise freedom, individualism, and peace as the highest values, Arab social structures—historically based on community, solidarity, and tribal ties—are destroyed by aggressive competitiveness and unexplained hostility. Said claims that according to Western thinkers, Arabs "naturally" live in the world of distrust and suspicion. Furthermore, the Arab need for revenge dominates all other values and is such an important aspect of their life that they are not able to confront this destructive sense of disgrace. Disunity, conflict-ridden nature and backwardness that characterize Arab societies inevitably lead to failures which make innovation and progress impossible.[6]

Furthermore, for centuries, the Orient world has been treated as a homogenous region, without taking into account its ethnic, religious, political, and economic social diversity. It is commonly ignored that there are twenty-two Arab countries in Asia and North Africa, from Morocco in the west and the Persian Gulf in the east. Another popular misconception about Arab countries, the Middle East in particular, is that they are all Muslim. This is an incorrect simplification, the reality being somewhat different:

> From the geographical definition, the Middle East is from Turkey in the north to Yemen in the south and from Egypt in the west to Iran in the east. This area contains three non-Arab states, Turkey, which is Turkish and Kurdish, Iran which is Persian, and Israel, the Jewish state. On the other hand, half of the 22 member states of the Arab League are located in North Africa and further south in northern Africa. Moreover, the Islamic world is wider than the above area because Islam is a dominant religion in most of northern Africa, Central and Southern Asia, Southeast Asia, and even parts of China. Indonesia, a non-Arab country in Southeast Asia, is the world's most populous Muslim country.[7]

In fact, as Jack G. Shaheen underlines, "only 12% of the world's Muslims are Arabs."[8] Therefore, the popular conviction that all Arabs are

---

[5] Ibid., 76–78.
[6] Ibid., 89.
[7] Chung and Kleiner, "Common Racial Stereotypes," 4.
[8] Shaheen, *Reel Bad Arabs*, 4.

Muslims and all Muslims are Arabs is obviously an error or—more precisely—a simplification often made to sustain stereotypes and generalize about this diverse ethnic group.

Generally speaking, representation of the Orient in Western culture has had little to do with reality but—as Greta D. Little writes—"truth was less relevant than Western perceptions and beliefs. Whether true or false, these constructed realities became received knowledge."[9] This knowledge has influenced both literary and visual representations of Arab and Muslim cultures, which are often referred to as identical, and the media very often forget, also, that these two terms cannot be used interchangeably.

The misinterpretations of the Orient cultures described by Said were transferred onto Arab and Muslims diasporas in the United States, where they have had to struggle with stereotypes and confront racist attitudes.

Arab Americans

Arab Americans constitute an ethnic group made up of several waves of immigrants that began to arrive in the United States in the 19th century. As underlined above, they originate from twenty-two Arab countries in Asia and North Africa; however, despite national diversity, they "represent common linguistic, cultural, and political traditions."[10] It is very important to acknowledge that the majority of Arab Americans are Christians; however, "due to the steady increase of immigration since the 1950s, Arab Muslims represent the fastest growing, albeit still minority, segment of the Arab American community,"[11] which today constitutes about 40% of the whole Arab population in the US. The religious beliefs of Christian Arabs are not so different from those of European Americans, whereas Muslim Arabs in the United States

> have many more religious traditions and practices that are unique to their faith and may compete with prevailing American behavior and culture. The beliefs of Islam place importance on modesty, spurn inter-faith marriage, and disapprove of American standards of dating or gender integration. Religious practices that direct personal behavior—including the five-times-daily prayers, the month-long fast at Ramadan, beards for men, and the wearing of *hijab*—and that require accommodation in such places as work,

---

[9] Little, "Representing Arabs," 264.

[10] Samhan, *Who Are Arab Americans*, 1.

[11] Ibid.

> schools, and the military, make Muslims more visible than most religious minorities and often vulnerable to bigotry.[12]

The first wave of Christian settlers from Arab countries, similarly to immigrants from other parts of the world who arrived during the same period, confronted prejudice, ignorance and nativist sentiments. Studies on immigration waves to the United States confirm that all ethnic groups suffered similar discrimination.

At first the image of the Orient, as underlined above, was exotic and idealized, which of course did not preclude perceiving it as weak and disunited. After World War II, and especially since the 1970s, the perception of Arab and Muslim countries in the United States significantly worsened and the idealization started to change into a growing hostility. Simultaneously, the depiction of Islam and the Middle East was becoming increasingly negative. This was the result of several factors, which served to make relations with this region a political issue. Anti-Arab and anti-Muslim prejudice—which in fact overlaps the whole history of Orientalism—was strengthened first of all by the Arab-Israeli conflict, with the United States playing a strategic role and providing the military and political support to Israel.[13] The struggle between Arabs and Jews introduced a simplified distinction between "democratic, freedom and peace-loving Jews" and "bad, totalitarian and terrorist Arabs."[14] The second event was the oil crisis of the 1970s which led to a significant increase in gas prices that were experienced by the average American on a daily basis."[15] The oil crisis altered the dominant stereotype of the poor Arab and contributed to the new image of wealthy and greedy oil sheiks. The subsequent events in the 1980s and 1990s—the conflict in Afghanistan, the Iraqi invasion of Kuwait, the Gulf War, and finally the September 11 attacks—ultimately antagonized the already hostile attitudes and "combined to create misguided mindset, leading some Americans to believe all Arabs are terrorists and that Arabs do not value human life as much as we do."[16] All these factors had a significant impact not only on the public opinion, but also on American political actions. Furthermore, they certainly influenced the media coverage as well as film and television productions.

---

[12] Ibid.

[13] Ibid., 3.

[14] Said, *Orientalizm*, 63.

[15] Basara, "Dobry muzułmanin ratuje Amerykę."

[16] Shaheen, *Reel Bad Arabs*, 29.

Misinterpretations of Arab/Muslim Cultures: Negative Representations in Film and Television

In her article about depictions of Arabs in literature, Little distinguishes several misconceptions about Arabs, which have also permeated other media, including film and television.[17] All the stereotypes she identifies date back to the first orientalist misconceptions of the East and refer to an alleged lack of intelligence, the deviant and inferior nature of Arab mentality, as well as its inclination to aggressive behavior. Thus, it would seem that Arabs are lazy, dirty, ignorant, superstitious, silly, irrational, cruel and violent. Additionally, "writers charge Arabs of lacking ambition and repeatedly show the men gathered in cafes smoking their waterpipes."[18] Moreover, other "immanent" features of Arab and Muslim culture are underlined—they mistreat women and hate Christians. The first one refers to a dominant conviction that "Muslim women, especially those in the Arab world, are depicted as victims of neglect or exploitation with no meaningful role in life."[19] The second contributes to one of the most popular and harmful contemporary stereotypes which states that Islam is a violent religion with *jihad* against the West at its center: "Presentations and representations of Islam as a religion reduce it to rote recitation and unfounded superstition. The violence and irrationality are now centered in acts of terrorism. Arab hatred for Christians is now a more general antipathy to the West, targeting Israel and its allies especially. They no longer trade in slaves; they now deal in hostages."[20]

Bearing in mind the changes which dominant depictions of Arabs and Muslims underwent after World War II, it can be stated without exaggeration that they are essentially seen as villains and Islam is reduced to aggressive ideology directed against the West, especially against American culture. Muslim religion is linked with "male supremacy, holy war, and acts of terror, depicting Arab Muslims as hostile alien intruders, and as lecherous, oily sheiks intent on using nuclear weapons."[21] Said comments on this situation as follows:

> So far as the United States seems to be concerned, it is only a slight overstatement to say that Moslems and Arabs are essentially seen as either oil suppliers or potential terrorists. Very little of the detail, the human

---

[17] Little, "Representing Arabs," 264–66.

[18] Ibid., 267.

[19] Ibid.

[20] Ibid.

[21] Shaheen, *Reel Bad Arabs*, 9.

> density, the passion of Arab-Moslem life has entered the awareness of even those people whose profession it is to report the Arab world. What we have instead is a series of crude, essentialized caricatures of the Islamic world presented in such a way as to make that world vulnerable to military aggression.[22]

Said also claims that an important aspect of the postmodern world of the electronic era is the tendency to strengthen stereotypes concerning Oriental cultures. This is the process wherein film and television industries—considered the most powerful mass media—play a key role by distributing standardized and simplified images.[23]

In his book *Reel Bad Arabs: How Hollywood Vilifies a People* Lebanese-American media researcher Jack G. Shaheen analyzed over 900 American movies in the context of Arab representation:

> What is an Arab? In countless films, Hollywood alleges the answer: Arabs are brute murderers, sleazy rapists, religious fanatics, oil-rich dimwits, and abusers of women . . . What do you see? Black beard, headdress, dark sunglasses. In the background—a limousine, harem maidens, oil wells, camels. Or perhaps he is brandishing an automatic weapon, crazy hate in his eyes and Allah on his lips.[24]

The depiction of women is equally stereotypical; what is different is that their representations are scarce, often reduced to "belly dancers and harem girls"[25] as well as "Black magic vamps, or enchantresses 'possessed of devils.'"[26]

The new negative stereotypes identified both by Said and Shaheen "emerged in and permeated advertising, television, and movies, particularly those of the nefarious oil sheik and the terrorist. The Arab as villain has been a favorite scapegoat of popular American culture."[27] Among these pejorative cliches that are systemically sustained by the commercial culture, Mahboub E. Hashem finds "depictions of Arabs as Muslims, fabulously wealthy, barbaric, uncultured, backward, terrorists, murderers, oil sheiks, Bedouins, desert dwellers, and sex maniacs with harems and a penchant for Western women."[28] It is very often the case that

---

[22] Said, "Islam Through Western Eyes."
[23] Said, *Orientalizm*, 62.
[24] Shaheen, *Reel Bad Arabs*, 2.
[25] Wingfield and Karaman, "Arab Stereotypes," 132.
[26] Shaheen, *Reel Bad Arabs*, 22–23.
[27] Samhan, *Who Are Arab Americans*, 3.
[28] Quoted in Little, "Representing Arabs," 261.

an oil sheik's onscreen presence is connected with kidnapping white women.[29]

Contemporarily, I think, the most damaging images are those of Arabs and Muslims as terrorists: "Today's films present anti-Christian, anti-Jewish Arab potentates perched atop missile bases, armed with nuclear weapons, plenty of oil, and oodles of cash. Using Islam to justify violence, today's reel mega-rich hedonists pose a much greater threat to the West, to Israel, and to fellow Arabs than did their predecessors."[30] The stereotype of an Arab-terrorist was so deeply rooted in social consciousness that while covering the Oklahoma City bombing in 1995, the media and public opinion were convinced that the terrorists were Middle Eastern fundamentalists: "[W]ithin minutes of the Oklahoma bombing, news reports were insinuating that the bombing was an act of terrorists. . . . Steven Emerson, a terrorist expert, told viewers not to believe Islamic groups when they denied involvement."[31] Finally, it turned out that the terrorist was an American. However, the impression of Arabs' alleged responsibility for the attack etched itself permanently in public memory. What is more, "speculative reporting, combined with decades of harmful stereotyping, resulted in more than 300 hate crimes against them."[32] Therefore, it is clear that depicting Arab Americans in such a negative and offensive manner leads to associating them with, or blaming them for, the acts of a small group of extremists who share their ethnicity or religion.

Considering the above mentioned misconceptions about Arabs and Muslims, it was a truly new experience to watch the television series *Sleeper Cell* which—as I read it—offers a different point of view and questions the common cliches.

Muslim and Arab Americans on *Sleeper Cell*

Sleeper Cell, produced by a private cable station Showtime, was broadcast in the fall of 2005. Although it evoked heated debate and met some criticism, in my opinion, the series is a generally successful attempt to challenge several ethnic and religious stereotypes. Undoubtedly, it is the first series in television history wherein a Muslim American is the main hero. *Sleeper Cell* tells the story of an FBI undercover agent who tries to infiltrate a small terrorist group in Los Angeles which is preparing a

---

[29] Wingfield and Karaman, "Arab Stereotypes," 132.
[30] Shaheen, *Reel Bad Arab*, 21.
[31] Chung and Kleiner, "Common Racial Stereotypes," 5.
[32] Shaheen, *Reel Bad Arabs*, 7.

massive attack on a public stadium. Seemingly, the plot sounds unoriginal and unsurprising, certainly typical for a detective/action production. However, the perspective changes when we discover the agent's identity. Darwyn Al-Sayeed, alias Al-Hakim (Michael Ealy) is an African American and a practicing Muslim. This interesting reversal of traditionally fixed roles challenges the common conviction that all Muslims are Arabs, as well as the stereotype that Muslims are terrorists or at least support them. The producers declared that they wanted to meet expectations of the six-million Muslim community living in the United States and expected that *Sleeper Cell* would help to question the dominant stereotypes about Muslims as well as reduce prejudice, especially towards Arabs. The series producer, Ethan Reiff, says he believed that four years after the Al-Kaida attacks on New York, Americans were ready for such a production. He underlines that so far both the terrorists and the CIA or FBI agents hunting them have been represented in popular culture as comic book heroes, whereas the show's creators decided to design the characters in a more realistic way.[33]

I believe the series' producers purposefully chose an African American for the main hero—first of all, this decision acknowledges the increasing popularity of Islam among Blacks since the 1960s when Malcolm X's teachings gained popularity. Secondly, it tries to persuade the audience that associating Islam exclusively with the Arab world is simply incorrect.

Interestingly enough, the cell's members, who are planning the terrorist attack on Dodger Stadium in Los Angeles, are a nationally and ethnically diverse group. It is clear that the director Clark Johnson's intention was not to present the *jihad* exclusively along ethnic and religious lines (as so far it has been commonly perceived as a conflict between the Muslim Arabs and the Christian West) but as a conflict within the Islamic world. In order to avoid this simplified understanding of tensions between Arab communities and other Americans, the scriptwriters depicted extremists as members of different countries: France, Albania, Saudi Arabia, and the United States. The cell consists of a French skinhead, Christian (Alex Nesic), who converted to Islam; an Albanian Muslim from Sarajevo, Ilija (Henri Lubatti), a veteran of the conflict in former Yugoslavia, whose family was killed by Orthodox Serbs; a young, blond-haired and blue-eyed white American, Tommy (Blake Shields), whose fascination with Islamic fundamentalism is clearly an expression of a rebellious attitude towards his upper-class, leftist, intellectual parents who are professors at Berkeley University. And finally

---

[33] Basara, "Dobry muzułmanin ratuje Amerykę."

there is Faris Al-Farik (Oded Fehr) from Saudi Arabia—the only Arab in the whole team and at the same time its leader. Al-Farik has a very interesting alias: "In Los Angeles, he goes by the name Yossi, posing as a Jewish businessman who coaches Zionist Little League."[34] When we add the African-American agent Al-Sayeed to this group, the cell seems to be far from the stereotypical image of Arab terrorists.

The series producers attempt to depict the cell's members as common people, often perfectly assimilated American citizens. They are "killing machines with a human face," as Stanley calls them.[35] The slogan which advertised the series was: "Friends. Neighbors. Husbands. Terrorists." Numerous scenes introduce viewers into the private life of fundamentalists in order to emphasize how ordinary their everyday life is, for example when they meet to talk over their terrorist plans at a Muslim girl's birthday party or a bowling club. Ilja works as a chemistry teacher in a high school where he has to deal with lazy students. Thus, on the surface, they are normal citizens eating American doughnuts, drinking Starbucks coffee, and listening to American hip-hop music. They are often well-educated people who have made their choice for certain reasons, so the question *why* they have become attracted to the *jihad* idea is clearly stated. Of course the producers do not try to legitimize or justify their actions, but rather to understand their motivations.

The main character, Darwyn, not only is a Black Muslim trying to avert a possible repetition of 9/11, he also has to confront the prejudice within the Federal Bureau of Investigation.[36] His struggle with deeply rooted and very often unconscious misconceptions concerning his religion, which are widespread among his colleagues at work, is sometimes as difficult as working out the terrorists' plans. Thus, on the one hand, Darwyn has to infiltrate an international organization of Muslim fanatics and convince them that he is a supporter of *jihad* and a fundamentalist ready to die for the cause. On the other hand, he advocates Islam as a religion of peace and dialogue—this aspect is clearly stated and very often underlined. In numerous conversations with his supervisors, he explains what Allah means to American Muslims. This certainly is something new on American television because, normally, "producers fail to recognize that 'Allah' is Arabic for God, that when they pray, Arab Christians and Muslims, use the word 'Allah' . . . [and] mislead viewers, wrongly

[34] Stanley, "Among Terrorists."
[35] Ibid.
[36] Basara, "Dobry muzułmanin ratuje Amerykę."

implying that devout Arab Muslims do not worship the 'true God' of the Christians and Jews, but some tribal deity."[37]

Sleeper Cell covers a variety of issues such as sources of financing Islamic terrorism (for example drug smuggling), the relations with similar organizations in Indonesia or Afghanistan, or the war in Iraq. However, equally important for the plot are themes connected with the incompetence of Darwyn's supervisors in FBI as well as their anti-Muslim stereotypes and ignorance concerning Islam.[38] Some room is also given to everyday prejudice and common beliefs about the Muslim religion in American society.

From my point of view the most important advantage of this series, in the context of contemporary Muslim-American relations, is the debate about the internal problems of the Muslim community and Koran misinterpretations by Islam worshippers themselves. This issue is directly addressed in the episode *The Scholar*, shot in a Los Angeles mosque and devoted to a religious leader from Yemen, Zayd Abdal Malik (Marc Casabani), who explains the true meaning of the Koran. He begins his teachings by asking a question: "What is the greatest threat facing Islam today?" The worshippers give different answers: American aggression in the Middle East or Zionist occupation of Palestine. The scholar responds: "These are merely symptoms. The greater disease, the real threat is in our hearts. Instead of allowing Islam to transform us, some Muslims are trying to transform Islam to serve their political agenda. They are using the label Islam to kill, to murder innocent people. God does not change the condition of a people until they change what is in themselves. How can we end the oppression of the Muslims by resorting to even worse behavior?" He emphasizes that the real "holy war" is not killing the enemies but a struggle with one's own temptations: "What is the greatest *jihad*? The greatest *jihad* is to battle your own soul, to fight the evil within yourself." Although the scholar is killed in the episode's final scene by one of the extremist, Christian, the important message is delivered to the audience by Darwyn who explains to his FBI supervisor: "This isn't just a war on terror, it's a war within Islam and people like Abdal Malik are the only ones that can help us win."

It is not only in the case of the terrorists that superficial representation is avoided; in *Sleeper Cell* there are some—although not very numerous—images which depict Muslim communities as assimilated to the dominant culture, albeit still preserving their traditions, as educated people who have

---

[37] Shaheen, *Reel Bad Arabs*, 11.
[38] Basara, "Dobry muzułmanin ratuje Amerykę."

nothing to do with Islamic fanatics. In this sense—as Kamran Pasha, one of the scriptwriters and second generation Pakistani, hopes—*Sleeper Cell* plays a similar role to that which is often ascribed to the comedy series *The Cosby Show*. Its enormous success contributed to negotiating dominant stereotypes of African Americans and made white Americans change the perception of Blacks, who started to be seen as fathers, mothers, regular citizens, and not just as an undifferentiated crowd of villains.[39]

The series also indirectly touches upon the problem of negative media coverage—we see Muslim communities televised only when a terrorist attack or another crime is committed. The role of television in propagating negative representations is visible in a scene wherein the evening news cover the murder of a Pakistani girl. The murder was commissioned by her uncle after she had been caught dating a non-Muslim man, not wearing her *hijab*, and breaking the fundamental religious rules (e.g. drinking alcohol). When the media concentrate only on such isolated and drastic events, they sustain the shocking image of Islam as a cruel and merciless religion that allows the killing of innocent people in the name of Allah. By doing so it also underlines all negative characteristics of the Orient described by Said: backwardness, vindictiveness, disrespect for women's rights, etc.

Although in general critics valued the performance of the protagonists and the producers' ambition to present Muslim communities in a different, more favorable light, they also indicated that *Sleeper Cell* failed to present or comment on the Israeli-Palestinian conflict. It also did not succeed in emphasizing that the lack of democratic reforms in Arab countries ruled by oil oligarchs fuels the fundamentalist notion of *jihad*.[40] Other commentators were also struck by an unusual image of the American metropolis, Los Angeles, which is depicted in a "decadent style known from the gangster film noir."[41] As Stanley notes: "The Islamic moles live in Los Angeles that is a nightmare tableau of why they hate us: seedy strip clubs, doughnut shops, 24-hour malls, S.U.V.'s neons and mindless consumerism."[42] This, however, is in my opinion one of the advantages of the series—the image of the US is far from the idealistic American Dream as it presents not only expensive upper-class neighborhoods but also slums and shabby parts of the city. Ibrahim Hooper, the spokesman for the Council on American-Islamic Relations in Washington, acknowledges *Sleeper Cell*'s positive potential in its attempt to demythologize Islam.

[39] Basara, "Dobry muzułmanin ratuje Amerykę."
[40] Ibid.
[41] Ibid.
[42] Stanley, "Among Terrorists."

Nonetheless, he also expresses some fears that viewers will remember shocking scenes of cruelty motivated by religious fanaticism rather than the positive Muslim hero.[43]

What I consider the biggest failure of the series is the striking absence of Arab and Muslim women—an issue that was generally ignored by the critics. Although they seem equally important, an improvement in representing women did not go hand in hand with a more complex depiction of men and Islam. On both sides, the war on terror seems to be an entirely male issue; women are just speechless witnesses or victims of men's actions. *Sleeper Cell* only confirms what Shaheen wrote about film representation of Arab women:

> Her mute on-screen non-behavior and black-cloaked costume serve to alienate the Arab woman from her international sister, and vice versa. Not only do the reel Arab women never speak, but they are never in the work place, functioning as doctors, computer specialists, school teachers, print and broadcast journalists, or as successful, well-rounded electric or domestic engineers.[44]

Some critical voices were justified and not exaggerated. Nevertheless, it must be underlined that *Sleeper Cell* creates a totally new image of the US in the era of war on terror and of Muslim-American communities. Although not entirely successful and criticized for its inaccuracy in presenting the home front fight with terrorism and the way the cell operates, the series is a significant step towards challenging the dominant perception of Arabs and Muslims by contemporary American society. Some of the issues the first season missed were developed in the second part of *Sleeper Cell*, which continues to be very critical of US politics: "It doesn't condone jihad or the hate that fuels it, but it tries to show why they hate us, and in doing so goes further than any other post-9/11 drama on American television. . . . Terrorists on *Sleeper Cell* are evil, but the United States is neither innocent nor blameless and carries the seed of its own decline."[45]

Conclusions

Mediated and, in this context, televised American culture is still awaiting the moment of true confrontation with prejudice towards Muslims and Arabs as well as with the notion of the war on terror and its

---

[43] Basara, "Dobry muzułmanin ratuje Amerykę."

[44] Shaheen, *Reel Bad Arabs*, 23.

[45] Stanley, "Inside Terrorist's Tangled Web."

strongly populist, emotional and negative language used both by politicians and the media. It is always the case, not only in the United States, that such negative attitudes towards the Other increase when the society undergoes a crisis. Unsurprisingly, they intensified after the 9/11 attacks on the World Trade Center, when nationalistic and nativist voices called for building walls, tightening borders, deporting illegal "aliens," banning the inflow of new immigrants, as well as limiting their rights.

Researchers underline that in comparison with other ethnic minorities in the United States, for example African Americans, there has been little progress in depicting Arab culture, especially Arab Muslims: "The media had not extended a second look to Arabs, who continued to be portrayed as wealthy, hedonistic princes; warlike, anti-American fanatics; as barbaric and repressive polygamists engaged in White slavery."[46] The only way to overcome these simplified stereotypes is to provide a greater diversity of images as "ethnic stereotypes are especially harmful in the absence of positive ethnic images."[47] Shaheen observes that Arabs are hardly ever seen as ordinary people, practicing law, driving taxis, singing lullabies or healing the sick.[48] Generally speaking, Arab and Muslim American communities are not provided with the representation of Everyman: "Missing from the vast majority of scenarios are images of ordinary Arab men, women, and children, living ordinary lives. Movies fail to project exchanges between friends, social and family events."[49]

In the "Preface" to the 2003 edition of *Orientalism* Said claims that misinterpretations of the Middle East, Arabs and Islam have remained a source of distortion and controversy, as well as military actions.[50] He also contends that the common understanding of the East and its cultures has not changed for the better in the United States as 9/11 led to a radicalization of attitudes, dissemination of degrading generalizations and slogans as well as to a further exclusion of ethnic/religious Others from the media discourse.[51]

Keeping in mind all these pessimistic comments, *Sleeper Cell* can be considered a small step in a good direction, which may initiate a discussion about different values and misunderstandings between the two cultures. For the first time in American television's history, mediated culture turned also to Muslim communities who condemn fanatics and

---

[46] Little, "Representing Arabs," 261.
[47] Wingfield and Karaman, "Arab Stereotypes," 133.
[48] Shaheen, "Media Image of Arabs."
[49] Shaheen, *Reel Bad Arab*, 15.
[50] Said, *Orientalizm*, 10.
[51] Ibid., 12.

Islamic fundamentalists, who struggle to coexist in American society and who silently hope that their image in the media finally alters.[52]

Works Cited

Basara, Zbigniew. "Dobry muzułmanin ratuje Amerykę." *Gazeta Wyborcza*, December 6, 2005.

Chung, Szu-Hsien, and Brian H. Kleiner. "Common Racial Stereotypes." *Equal Opportunities International* 22, no. 3 (2003): 1–9.

Little, Greta D. "Representing Arabs: Reliance on the Past." In *Cultural Diversity and the U.S. Media*, edited by Yahya R. Kamalipour and Theresa Carilli, 261–71. New York: State University of New York, 1998.

Said, Edward W. "Islam Through Western Eyes." *Nation*, April 26, 1980.

—. *Orientalizm*. Translated by Monika Wyrwas-Wiśniewska. Poznań: Zysk i S-ka, 2005.

Samhan, Helen Hatab. *Who Are Arab Americans?* Washington, DC: Arab American Institute Foundation, 2001. http://www.aaiusa.org/page/file/21b02cde94d4307c47_jsnmvy5dd.pdf/WhoAreArabAmericans.pdf (accessed March 17, 2009).

Shaheen, Jack G. "The Media Image of Arabs." *Newsweek*, February 29, 1998.

—. *Reel Bad Arabs: How Hollywood Vilifies a People*. New York: Olive Branch Press, 2001.

—. *The TV Arab*. Bowling Green, OH: Bowling Green State University Press, 1984.

Stanley, Alessandra. "Among Terrorists, a Hero is Lurking." *New York Times*, December 2, 2005.

—. "Inside Terrorist's Tangled Web." *New York Times*, December 8, 2006.

Wingfield, Marvin, and Bushra Karaman. "Arab Stereotypes and American Educators." *Social Studies and the Young Learner* 7, no. 4 (March–April 1995): 132–36.

---

[52] Basara, "Dobry muzułmanin ratuje Amerykę."

THE MESSAGE IS THE MASSAGE: RETHINKING MARSHALL MCLUHAN'S THEORIES IN THE AGE OF NEW MEDIA

MIROSŁAW ALEKSANDER MIERNIK

Marshall McLuhan is best remembered today for his statement that "the medium is the message." As is the case with most of the catch-phrases and one-liners for which he was known, this one also caused, and still causes, various problems with its interpretation. The version upon which most agree is that by this phrase he meant that any medium shapes a person's cognitive abilities. This is in accordance with his theory that all tools used by human beings are extensions of the body. Just as McLuhan sees an axe as an extension of the human hand or the wheel as an extension of the foot, he claims the media to be the extensions of the central nervous system.

In McLuhan's view, the electronic media were to cause as significant a revolution in the age of print, as the printing press caused in the age of tribalism, essentially by changing the patterns in which people think. Although his theories gained popularity in the sixties, a vast majority of them could neither be proven nor rejected empirically, which eventually led to the diminishing of his popularity, even to the extent that nowadays he is very often considered an intellectual charlatan.

The question whether the "medium is the message" has been widely discussed; however, Justin Lewis noted that communication needs form, citing the example of news stories which very often are easily forgotten, which he attributes to the lack of narrative structure.[1] Jean Baudrillard went even further in his theory of the media's implosion: assuming that the medium is the message he claimed that both (particularly in relation to electronic media) are non-existent in the modern world. He notices that it is the message which may lend credibility to the medium, and that in a world of hyperreality the dependence between them is circular. This is a two-way connection: if the message is distorted, so is the medium; if the

[1] Lewis, "Absence of Narrative."

medium is distorted, the same applies to the message.[2] Baudrillard points out that the media is filled with a vast amount of useless information, entropy; as such, it incapacitates people's communication abilities.

McLuhan was a stark technological determinist; he believed that the tools we use change our "ways of thinking." He claimed that the electronic media would usher in an era of the "global village," a new tribalism. Although he emphasized the fact that he was not an advocate of these changes and claimed only to be "predicting" those which would take place in the future, he does seem to be rather optimistic about them. His "predictions" often seemed extremely far-fetched, as is the case with his views regarding the dissolution of the USA and other countries (due to racial incompatibility and the resurgence of ethnic identity), the disappearance of large cities, such as New York, Chicago and Los Angeles (as cities are also an extension of man, they will "wither" when entering a new media age), and the end of automobiles (which he thought will shortly be replaced by anti-gravitational vehicles).[3] These were supposed to be the effects of the electronic media; despite the fact that they are still closer to science fiction than to reality, there are other matters which can be investigated. McLuhan argues, for instance, that the global village created by the media would be highly vulnerable to an international "media shock" caused by certain events. Such events would surely include President Kennedy's assassination, as well as the terrorist attacks of 9/11. However, other events of this kind are much less significant on the global scale. The death of celebrity and second-grade movie actress Anna Nicole Smith gained major media attention in the US, but the importance of this fact as a news item waned elsewhere. The O. J. Simpson car chase may not seem as significant today as it was considered when it happened, and while the Columbine high school massacre still seems shocking, albeit slowly drifting toward the forgotten, other similar tragedies receive significantly less media attention.

When McLuhan died in 1981, the TCP/IP protocol, the base protocol used by the Internet, was in its early test phases. The mid-nineties saw the Internet becoming an increasingly popular medium. Internet World Stats site cites that in the first half of the year 2007 16.9% of the world's population could be classified as Internet users (Europe: 38.9%, North America: 69.7%, Australia and Oceania: 53.5%; note that Europe includes many countries which are underdeveloped according to the standards of

---

[2] Baudrillard, *Simulacra and Simulation*, 79–86.

[3] McLuhan, "Playboy Interview," 233–69.

the US or Australia).[4] This result is impressive once one takes into account the fact that this is a new medium, which needs a relatively expensive infrastructure, both on the part of the provider and the user.

Having recently passed the threshold of the 21st century, we have been living in a world of electronic media for a fairly significant period of time: experiments with the electric telegraph were conducted since the very beginning of the 19th century, and it entered commercial use in 1837. However, the major change was introduced when the electronic media became mass media in the early twentieth century. McLuhan's predictions made in the sixties still remain strongly hypothetical, but a look upon mass media (and particularly the electronic mass media) may give us some information as to whether they are developing according to his views.

One could argue that the printing press put an end to tribalism by promoting such elements as individualism, the scientific method, industrialism, or world trade. It is certain that the printing press facilitated the development of humankind by giving us the means to convey our thoughts and discoveries to others in an unadulterated form; this was impossible both in the era of oral-aural communication and when documents were written by monastic scribes. The printing press minimized this problem, but we can only guess if it changed our "ways of thinking." Nevertheless, one cannot disregard other effects it had on people, the most important ones being the popularization of literacy and knowledge, and along with that, massive cultural changes which helped societies develop greatly since the printing press was originally invented.

When Europe and America were to be connected by telegraph, Henry David Thoreau was one of the skeptics who believed that "the first news that will leak through into the broad flapping American ear will be that Princess Adelaide has the whooping cough."[5] He predicted an essential change ushered in by the popularization of the media—the substantial loss in the quality of its content. This was visible as newspapers became increasingly popular in the nineteenth century.

Electronic mass media pose an interesting advancement. McLuhan prophesied the end of the printing press with the advent of the electronic media. Though it may have come as a surprise to many of his enthusiasts, the electronic media seem to indulge the printing press. Today, most newspapers in the Western world are available online, free of charge. Still, "printed" newspapers remain popular, and although many have shown

---

[4] Internet World Stats, "World Internet Usage Statistics News and Population Statistics," Miniwatts Marketing Group, http://www.Internetworldstats.com/stats .htm.

[5] Thoreau, *Walden, or Life in the Woods*, 54.

diminishing sales, sometimes to the extent that certain publications, once popular, are being shut down (*Life* magazine may serve as a significant example here), others retain their high sales. Yet, since 1998 (previous data is incompatible due to a change of research methods) daily newspaper readership amongst adults has fallen from 58.6% to 48.4%,[6] while the popularity of online editions of dailies has risen. In terms of individual newspapers, some did show a rise in sales in 2006, but only a miniscule one, as the examples of *USA Today* (0.09%) and the *New York Times* (0.5%) illustrate. Those newspapers which experienced a drop in sales were the *Wall Street Journal* (1%) *Los Angeles Times* (5.4%), *The Washington Post* (3.7%). *The San Francisco Chronicle* experienced the most significant decrease of sales (15.6%). This trend has been present since the eighties, when alternatives to the printed press appeared, both in the form of news-themed television channels (such as CNN) and the Internet.

Despite these changes, there are no signs that the newspaper in its printed form will disappear in the nearest future. It could be claimed that many people simply prefer to read "paper" editions of newspapers out of habit or that their readers might be people who do not use the Internet. It should also be noted that the new media are not as mobile as newspapers are, and the technologies which enable their mobility are still out of reach for most of the world's population.

This does not mean that Internet users shun the printed word: statistically, an Internet user both reads and purchases more books than a person who does not use this medium. Despite the advance of the Internet, newspapers maintain a strong market position, including stock markets, in spite of the fact that the electronic media have been around for decades. There are certain facts that complicate this picture of peaceful coexistence of the old and new media. Digital cameras, for instance, outsold traditional cameras in the US as early as 2003,[7] but it also should be kept in mind that they are much cheaper to use. Besides, there are various other advantages to digital cameras.

When set against McLuhan's theory of hot and cold media, the computer seems to be a lukewarm medium, regardless of how one interprets this obscure distinction. One may use the computer as a typewriter, a newspaper, watch a movie on it, play a computer game, browse through the catalogue of a university library, look through the

---

[6] Newspaper Association of America, "Daily Newspaper Readership Trend—Total Adults (1998–2007)," http://www.naa.org/marketscope/pdfs/Daily-Trends-Total Pop-98-07.pdf.

[7] Davies, "Kodak to Stop."

pictures one took during a holiday or listen to a radio station. Unlike television, the computer does not favor one method of communication over others. Despite teletext and news crawl, television has always been a medium of sound and vision—or a tribal one, to use McLuhan's own term. It even started to offer certain possibilities of interaction, such as text message competitions or shows where the viewers can vote for their favorite contestants/characters and to a degree influence the programs; these practices, however, might seem minimalist by the standards of the Internet. Modern media technology allows users to focus on only one means of communication, although most users rely on a reasonably balanced mixture. This would imply that the Internet is not a medium; it rather is a juxtaposition of several other media, closely intertwined and connected, but at the same time separate and autonomous, depending on the given case.

Many people associate the concept of the global village with the Internet. Although limited in its use, the Internet is available in practically all the countries of the world (although there are instances where only 0.03% of the population use it, as is the case with Liberia). Still, it is important to note that the Internet is strongly dominated by American culture, and many users of this medium react to it, whether by embracing it, praising the tenets of American democracy and individual expression, or the opposite—by rejecting it.

When scrutinizing the media and its contents, it is surprising that these two often seem fairly independent of one another; this puts aside the hypothesis of the medium being the message or massage which shapes the human being. An initial idea one might get is that the contents of the media are the most important elements in such cases, but the message/text is a body which needs interpretation. This interpretation, according to post-structural theory, lies in the domain of the reader. Nonetheless, the author does play a certain role in how his work will be read through its form. A given form may significantly change the bias of a text, owing to many elements.

There is a quote attributed to Napoleon Bonaparte which may serve as an example: "A man does not have himself killed for a half-pence a day or for a petty distinction; you must speak to the soul in order to electrify him." The history of humankind has known many speakers with high rhetorical skills. Still, somehow, in the age of the printing press, when the radio was still a relatively rare device, Adolph Hitler managed to electrify a whole nation in a way which allowed him to wage one of the most brutal conflicts in the history of humanity. Neil Postman gives the example of the Lincoln-Douglas debates, lasting for hours, which were observed carefully by many people, and claims that in today's world it would be impossible for large groups of people to participate in such an event. Postman describes a debate which lasted about three hours,[8] arguing that such public discussions were a display of the country's high political culture. However, in the middle of the 19th century this was also, and perhaps primarily, a form of entertainment, which has been replaced by other forms in modern times, hence the fall of demand for debates lasting many hours. Presidential debates are still an element of American political life, but their form has changed significantly. Although many would be eager to claim that this is due to the shortening of the human attention span, movies and theater plays which last over three hours and still manage to attain commercial success would counter these arguments.

A medium may spread the skills needed for its use. The growing popularity of newspapers greatly contributed to literacy,[9] and computers spread "computer literacy" in a similar way. The medium also places constrictions on the form that the message assumes in it. Due to technological progress this factor may vary, the best example being television, which through teletext can offer, for instance, subtitles for those with impaired hearing. The form has direct effect on the matter, and the medium's or media's effect on a person has to be conveyed by the form, which has to be accustomed both to the medium and the matter. There are also instances in which different media influence each other. The computer and the Internet, for example, affected other media significantly. The initial tendency was to focus on mass audience participation, but on the whole, television rather extended into the Internet than made attempts to emulate it, particularly trying to market their programs on-line through the use of additional materials, biographies of the actors and the characters they portray, additional information on popular programs, etc. Yet various media have influenced each other for a significant period of time. The effect of the printing press is obvious, particularly in the experiments leading us closer to e-paper, a portable, elastic, thin device the size of an A4 sheet of paper, which will have the capacity to store many books in its memory and display them in order to allow the portability of electronic books, which has been significantly challenged up to this day due to a number of factors. The influence of television and the radio is also obvious in computers. In the recent years it may be said we have undergone a visual revolution in the media. Introduced by television, it has affected other media: newspapers, particularly tabloids, have employed an

[8] Postman, *Amusing Ourselves to Death*, 44–47.
[9] Williams, *Long Revolution*, 173–213.

immense amount of visual material, to the extent that it clearly outweighs the amount of textual information. However, the lowering of quality is supposedly observed in every medium. A common opinion of this is that such lowbrow media affect us, effectively "dumbing down" the population. Although this theory is tempting, research proves something to the contrary (assuming that IQ is not a useless indicator): the average American's IQ level has gone up by 24 points since 1918.[10]

If there is a medium native to the Internet, it is hypertext. Little doubt remains that McLuhan would be very enthusiastic about it, as it seems to promote the non-linear fashion of processing information. Nevertheless, hypertext is very often criticized for this. The problem of "fascinated clicking"—the search for specific information being slowed down as a result of pursuing additional information or trivia unrelated to the initially sought knowledge—is often considered a nuisance by Internet users. It is an interesting element of many Web sites which are very strongly based on hypertext, and if one observes Internet browser windows with dozens of tabs open, it is a strong argument supporting a certain non-linear aspect of the Internet. Still, it is only an aspect, since whether the Internet will be used in this way is up to the user. Opponents of this claim might point out that it is generally a hi-tech equivalent of "doing many things at one time," though such experiments as the hypertext novel introduce a possibility of a highly non-linear setting. In spite of all these factors, the Internet still remains a primarily visual-textual medium. McLuhan saw television as a home for a similar non-linear form, but this factor is often criticized due to such claims as, for instance, news stories very often being easily forgotten, as discussed earlier in connection with Justin Lewis's theories.[11]

In this case we may ask ourselves: what is the message of the medium? Basing on the aforementioned, it is not unlikely that the medium plays an intermediate role in shaping the form and matter of a message, but there is no way of proving whether it affects us. For now, the global village, essentially a united states of mind, where people are shaped by technology, still exists solely in such subdomains of science fiction as cyberpunk. It is in the confines of this genre that terms such as "cyberspace" or "the Metaverse" first appeared, along with such ideas as grafting cybernetic implants into human beings, the most radical form of any extensions of man. It is in cyberpunk that the apocalypse is shown as the dissolution of neuroelectronic networks, after which humanity begins to die out, as only very few are able to survive without a continuous direct

---

[10] *Guardian*, "Dumbing up."
[11] Lewis, "Absence of Narrative."

connection to databanks containing the whole of humankind's progress and limitless amounts of information. The irony lies in the fact that the most important pieces of this genre—and, it should be noted, very highly revered amongst professional Internet enthusiasts—are of the Gutenberg variety, i.e. books.

Works Cited

Baudrillard, Jean. *Simulacra and Simulation.* Michigan: University of Michigan Press, 1994.

Davies, Catriona. "Kodak to Stop Making 35mm Cameras." *Guardian*, January 14, 2004. http://www.guardian.co.uk/technology/2004/jan/14/onlinesupplement.uknews.

Guardian. "Dumbing up." April 23, 2001. http://www.guardian.co.uk/dumb/story/0,,476898,00.html.

Lewis, Justin. "The Absence of Narrative: Boredom and the Residual Power of Television News." In *Television: Critical Concepts in Media and Cultural Studies*, edited by Toby Miller, 2:322–37. London: Routledge, 2003.

McLuhan, Marshall. "Playboy Interview: 'Marshall McLuhan—A Candid Conversation with the High Priest of Popcult and Metaphysician of Media.'" In *Essential McLuhan*, edited by Eric McLuhan and Frank Zingrone. New York: Basic Books, 1995.

Postman, Neil. *Amusing Ourselves to Death.* New York: Penguin, 1985.

Thoreau, Henry David. *Walden, or Life in the Woods*. New York: T. Y. Crowell, 1899.

Williams, Raymond. *The Long Revolution.* London: Hogarth Press, 1992.

NEW MEDIA: INTERNET EFFECTS ON AMERICAN NEWS REPORTING

RICHARD PROFOZICH

Internet Culture

The Internet is growing continuously. More and more people worldwide use it and access is becoming easier, cheaper, and more powerful. In the United States, whole cities are turning into wireless sites, allowing anyone with a computer to access the Web anywhere, anytime, for free. Yet, the Internet is just beginning to grow, and it is expanding into the fabric of everyday life. Its effects are beginning to be felt everywhere, changing ways we shop, rent or buy a house or car, communicate with friends or family, and receive information and news. Katz explains that "the Web is the central reality—economic, social, cultural, political—for younger generations, whether we like it or not, whether we think it is a good thing or not. I don't think the Web is supplanting culture; it is creating a new culture."[1]

This is particularly true as far as journalism and news reporting are concerned, especially in the United States. The Internet and World Wide Web are fundamentally affecting the way in which news is being gathered, presented, and received. In recent years, Americans have witnessed the creation of New Media, as the old-style media transformed into new, high-tech, high speed, ever-changing, always available, interactive versions of themselves. Much of the change has to do with the digitalization of information, making it much easier and cheaper to transport data anywhere at any time. We have witnessed the creation of what has been termed online journalism, civic journalism, citizen journalism, and the virtual newsroom.

The result, says Pavlic, is the transformation of journalistic storytelling from an omnidirectional flow controlled by a news organization, to a

---

[1] Katz, "Future is the Net," 15.

multidirectional flow controlled in large part by the audience. The change is dramatic and is part of the new culture being created by the Internet. This engenders a myriad of opportunities for New Media journalists, professional or amateur, and dictates major changes in the news industry. "Newspapers are an editor's medium," Pavlic goes on to explain, "and . . . broadcasting is a producer's medium . . . the Internet is a journalist's medium. The Internet not only embraces all the capabilities of the older media (text, images, graphics, animation, audio, video, real-time delivery) but also offers a broad spectrum of new capabilities, including interactivity, on-demand access, user control, and customization."[2] This means that a fundamental change in journalism has occurred, what Pavlic calls "the storytelling transformation of journalism" in the online environment to become what he calls "contextualized journalism." This new style journalism has five characteristics: breadth of communication modalities (audio, video, graphics, and animation), hypermedia (links to other sources), heightened audience involvement, dynamic content, and customization.[3]

Fredin and David make the point that "the fact that massive repositories of information are only a few mouse clicks away offers a richness to hypermedia that set it apart from traditional media." They add that at the same time hypermedia require or at least enable a much more active and participatory audience.[4] Online news content is much more fluid and dynamic. Users want their news on demand and in real time. "Cyberspace comes closer to the infinite than anything humans have ever invented," says Kebbel. It is "a Big Bang of thought and ideas and news and information that is remaking what we know and how we see and express ourselves."[5]

Journalists are just learning the characteristics of the Web, according to Stevens. As far as online news reporting goes, it is not just a matter of posting traditional newspaper or broadcast content on the Internet. "The Web is its own medium," Stevens points out. "For the first time, a communications medium mirrors life. Artificial constructs of 24 hour newspaper deadlines or multiple daily TV or radio deadlines are gone."[6] "The consensus among the online media professionals internationally . . . is that online journalism is definitely 'a breed apart.'"[7] Flavian and Gurrea

---

[2] Pavlic, *Journalism and New Media*, 3.
[3] Ibid., 4.
[4] Quoted in Katz, "Future is the Net," 16.
[5] Kebbel, "Challenge of Community Building," 27.
[6] Stevens, "Taking the Big Gulp," 66–67.
[7] Deuze, "Online Journalism."

make the point that "the use of these new systems is distinguished by the speed with which news reaches the reader, the low cost of distributing information, and the opportunity to establish more direct contact and to interact with users, who have come to play an important role in the design of journalistic offers."[8] The Web indeed has constructed its own culture. Other important elements of that culture are a constant updating of information, a sense of shared space and a belonging to a community, albeit a virtual one, and an expectation that information should be free. Also, as Craig Newmark, the developer of Craig's List (which has over five billion page views each month) explains, the Web has developed a culture of trust and self-policing.[9] However, this may prove to be one of the weaknesses of the Web, and an issue that journalists need to deal with. As Newmark goes on to say, "What's troubled me is that disinformation campaigns and information warfare appear to be amplified by the Internet. When the scent of bad information can taint a relationship built on trust, this poses problems for the Web."[10] Another important element of Web culture is the diversity and depth of information it makes available through its unlimited capacity to store information. As Katz points out, it has "infinite 'space' for words, photos, video and audio, multimedia capability, and interactivity."[11] Moreover, it offers the unique feature of free links to other sites. Online news, therefore, can supply much greater context than traditional news providers.

Online News Audience

In 2000, the number of Internet users in the United States reached 144 million, and the percentage of individuals and households who have access to an online communication service reached 55%.[12] According to a Pew Research Center for the People and the Press survey released August 9, 2007, roughly one quarter of Americans get their news primarily from online sources. As other studies have also verified, this audience tends to be younger and better educated than the public as a whole. Also, it tends to be richer. For instance, Bert Solivan, VP of news and information and general manager of Foxnews.com, reports that half of their online audience come from households making $75,000 a year or more. Users tend to be more critical also. "People who rely on the Internet as their

---

[8] Flavian and Gurrea, "Choice of Digital Newspapers," 232.
[9] Newmark, "Community Building on the Web," 25.
[10] Ibid., 26.
[11] Katz, "Future is the Net," 196.
[12] Pavlic, *Journalism and New Media*, 139, 169.

main news source express relatively unfavorable opinions of mainstream news sources and are among the most critical of press performance," says the Pew survey.[13] The major criticisms of traditional news organizations are that they lack empathy for those they report on, fail to "stand up for America," are politically biased, and try to cover up their mistakes. This is a bit ironic because, as Downie Jr. and Kaiser, among others, point out, the most popular online news sites are either affiliated with TV networks (CNN.com, MSNBC.com, ABCnews.com, CBSnews.com) or with the major independent or national newspapers (the *New York Times, Washington Post, USA Today, Wall Street Journal*).

Users turn to online news chiefly for breaking news stories about major events, such as the September 11, 2001 terrorist attacks on the United States, the December 2005 Asian tsunami disaster, or election results. "Big, breaking news stories draw the largest audiences to news sites," reports Katz.[14] *Washington Post* executive editor Michael Silberman echoes Katz's point, stating that "breaking news is our traffic driver."[15] Breaking news is the major way in which online news sites build readership, as users tend to come back. According to a study conducted by the Online Publishers Association, 66% of Internet users look for news and current events.[16] However, another main feature of the Internet culture which also draws users to online news sites is its infinite capacity to hold information from a myriad of sources. A 1998 Pew Center study shows that 77% of American users aged 18 to 29 value the diversity of perspectives they get from online news sources. This is something that can counter the mistrust of individual news sources. A 1998 Gallup poll shows that news reported on the Internet is viewed as "fair and impartial" by the largest percentage of adults in America, 53%.[17] Again, this could be partially due to the ability of online news to provide a broader context and variety of views on a story. Christopher explains that the Web demands context. Stories must be connected to other stories and filled with searchable databases and user-generated content. This is all part of what she calls "news as a conversation."[18] Online journalism, therefore, seems to have a dual function in reporting news and attracting online users: to provide easy access to fast-changing breaking news events, and to offer broad, in-depth coverage of issues users view as important.

---

[13] Pew Research Center for the People & the Press, "Internet News Audience."
[14] Katz, "Future is the Net," 196.
[15] Quoted in Katz, "Future is the Net," 197.
[16] Roberts, "Future of Evening News," 45.
[17] Pavlic, *Journalism and New Media*, 129.
[18] Christopher, "Journalists Learn," 8.

Noting that the newspaper industry has been dramatically affected by the development of the Internet and its use as a news medium, Flavian and Gurrea conducted a research study of online newspaper readers to gauge their reasons for choosing online news sources. They point out the lack of studies which explore the behavior patterns of online newspaper readers or the reasons which lead consumers to read either standard or online newspapers. Their 2004 research study hypothesized three reasons for reading newspapers, either traditional or online: 1) to search for specific information, 2) to search for updated news, and 3) for leisure and entertainment, to pass the time. Their results showed a significant relationship between the search for specific information and the search for updated news and reading news online. However, there was no significant correlation in the case of reading for leisure or entertainment. This study seems to verify the general belief that online users are more interested in news in the first place, and in keeping up-to-date on breaking events as well as on important issues and trends. It also supports the industry observation that, overall, news and current events make up the most popular content on the Web. As Pavlic explains, "most people go online to connect with the news of their community, whether geographic or of the mind."[19]

Added Value

If the Internet is indeed transforming culture, or even if it is a culture unto itself, with its own communication style and modality that threatens to "sweep away the existing media landscape, radically undermining the power of existing media gatekeepers to create a new, decentralized and more democratic media environment,"[20] the question remains: what is the added value of online journalism? What does it give to news consumers that is not available elsewhere? First of all, as Katz points out, the Web breaks down the model of a few people dispensing information to many, so that now many people are dispensing information to many. Professional journalists still perform the gatekeeping role and help to determine the validity of information presented on the Web, but now they have to share power.[21] Thus, the news business is transformed from a passive medium where viewers or readers are told what is important to a participatory one

[19] Pavlic, *Journalism and New Media*, 30.
[20] Naylor, Driver, and Cornford, "BBC Goes Online," 137.
[21] Katz, "Future is the Net," 16.

where they decide for themselves.[22] As Kansas and Gitlin acknowledge, "this democratization of information is not harmful, but rather it is empowering."[23] Commenting on a trend in American journalism toward attracting larger audiences at the expense of serious news coverage, Carey notes that "without the institutions or spirit of democracy, journalists are reduced to propagandists and entertainers."[24] The Web, therefore, may be the force that helps restore the democratic element essential to maintaining quality, meaningful news reporting.

Journalists debate whether or not the online news site just "cannibalizes" the core business of media organizations, taking news for online distribution before it appears in broadcast or in a newspaper. With the onset of the Internet as a major news source, the prevailing practice was for traditional news organizations to simply post the same stories on the Web. Newspaper Web sites generally looked like the paper edition posted online. Deuze points out that "the cannibalization discussion can be solved if the editors choose to provide added value to the audience particular to a site which offers content with internal hypertextuality and clear navigational interactivity; this is not the same audience as a print or broadcast product will have."[25] Noting that traditional broadcast or print news organizations sit on more information than they distribute to the public, Deuze goes on to say that "the value of [Internet] sites and this kind of online journalism can be found in their sheer archival capacity . . . the seemingly endless capacity of a Web server." Online sites then appeal to "those who want to seek out all details of the story or stories, including audio, video and written text."[26] Stevens adds that "the Web is infinitely deep and all points on it are connected. That means that stories no longer stand alone. They're embedded in a matrix."[27] She goes on to point out that in online journalism "the most energy comes from churn and from content contributed by users."[28]

With the realization that the Web has produced a different news culture, that the needs and expectations of online readers are different from those of traditional readers, and that stories posted online need to be different from their print or broadcast versions, news organizations have slowly developed a new storytelling paradigm—one that uses a

[22] Pope, "Network and Cable TV," 67.
[23] Kansas and Gitlin, "What's the Rush," 83.
[24] Carey, "Lawyers, Voyeurs and Vigilantes," 19.
[25] Deuze, "Online Journalism."
[26] Ibid., 9.
[27] Stevens, "Taking the Big Gulp," 67.
[28] Quoted in Christopher, "Journalists Learn," 6.

multimedia approach. This kind of contextualized storytelling is largely made possible by the technological advances of the Internet, the development of digitalized content, and the advancement of broadband access.

The new, contextualized mode of storytelling means that the audience need no longer accept any single point of view as the full story or complete truth. No longer is the audience passive. The Internet offers the power to talk back and to interact. Choosing what to read or watch and when is the simplest form of compelling interaction.[29] There is no reason to "dumb down" online news as the Web's technological capacities lend themselves to high-quality news offerings with depth, interactivity, and customized editing for different audiences. Pavlic discusses the many dimensions of contextualized storytelling and the value added by online news sources as a result of these technological developments. Inventions such as the omnidirectional camera, which offers a 360 degree transmission range, provide much greater access to information and context. Hyperlinks add a variety of perspectives and richness of content that appeal to more active and participatory audiences. The Web's infinite capacity to hold information allows full presentation of the problem; for instance, an entire Congressional report or court proceedings can easily be provided. News stories no longer need to focus on events at the exclusion of broad social, political, or economic trends or processes. All modalities of human communication are available on the Web, and audiences can respond, post their own content, or provide additional links. Furthermore, Internet news provides a new dimension of timeliness with continual updates and new developments; no longer is it necessary to wait until the next day to find out how a story is progressing. Also, Internet news is resistant to censorship, providing full coverage rather than screening out important content. This, indeed, is a new kind of storytelling. One that is richer, more involving, always changing.

Current News Sites

The most common form of online news is the mainstream news site which generally offers standard style reporting from the parent organization, usually a broadcast network or newspaper, usually with minimal and often moderated participatory communication. As Deuze explains, "this type of news site cannot be said to differ—in its approach to journalistic storytelling, news values, relationships with audiences—fundamentally

[29] Downie Jr. and Kaiser, *News About the News*, 216.

from journalism as it is practiced in print or broadcasting media."[30] The best sites belong to the biggest newspapers and networks. Most sites also provide possibilities for email, discussion sections, chats, web search engines, classified ads, online shopping links, and city guides to the places where they are headquartered. Some have email newsletters for subscribers who want tailored news packages. Most of the revenue for these online sites comes from site advertising, but that does not yet cover the cost of most Internet news sites. Big news organizations use the Web to defend brands and to spread their journalism, looking to future changes in audience behavior, whether or not it makes a profit.

The most visited news sites on the Web, however, are major portals like Yahoo! AOL, and MSN, used by many people as gateways to the Web. They do not create any journalism of their own but buy standardized news stories and summaries from the wire services, the Associated Press and Reuters. This is standard journalistic news reporting, offering speed, accuracy, and fairness in the objective tradition, what Downie Jr. and Kaiser call a "basic, bland diet of news and bulletins."[31] The wire services do not provide much depth, investigative or interpretative reporting, or go behind the scenes of events, but they stay on top of stories, updating them constantly. The portals also provide links to other areas, especially the mainstream news sites. They treat news as a commodity.

The smaller, independent online news sites often combine opinion journalism with a small amount of original reporting and a lot of chat. Most news here also comes from wire services, rewritten and updated, or summarized, with added headlines, graphics, video, or links. Staff also write personal commentary on news reported on TV or in newspapers. These sites will often provide dialogs between well-known writers or personalities about hot topics. Nevertheless, the sheer number of sites on the Internet has overwhelmed most Web information providers, and only a few independent sites have attracted many readers, let alone made a profit.

So far, the Internet has not become a significant source of original journalism, according to Downie Jr. and Kaiser. Rather it is an important news distribution medium for established news media. The authors claim that American consumers see the Web more as a user-friendly radio station than a serious provider of in-depth news. However, there is a strong and steadily growing demand for more serious journalism on the Internet. In consequence, traditional journalists have begun to see the Web as an opportunity to give readers more background, details, graphics, and videos

---

[30] Deuze, "Online Journalism."

[31] Downie Jr. and Kaiser, *News About the News*, 209.

that will not fit into newspapers or TV broadcasts. The idea is to blend the speed of broadcast news, the depth reporting of newspapers, and the unlimited capacity and interactivity of the Internet to offer a truly new kind of journalism.

Online Journalism

The Internet built an audience much faster than either radio or TV, and it seems logical to assume that it will continue to develop and attract users and audiences for news. It has the capacity to marry the advantages of radio, TV, and newspapers, which could make it much more powerful than all three. At the big mainstream news sites, more and more reporters produce stories for the Internet and participate in online chats and web videos. Holahan maintains that perhaps the biggest evidence of the change in mainstream journalism is "the redesigned *Wall Street Journal* and WSJ.com, launched on January 2, 2007. The print edition now concentrates on analysis stories, leaving the breaking news that once made up nearly half the newspaper for the online edition, which publishes throughout the day. Industry executives and analysts say the paper is going in the right direction and that other publications will have to follow suit or risk folding."[32] The *Wall Street Journal* is one of the most accessed mainstream news Web sites. Smaller newspapers are also focusing more and more on Web reporting, often directing their stories straight to the Web site, and later possibly into their print editions. The *Fort Myers News-Press*, part of the Gannett chain, has a fleet of mobile journalists, or "mojos" with ThinkPads, digital audio recorders, and digital still and video cameras who spend their time on the road looking for stories and filing them directly to the newspaper's Web site. "The chain's papers are redirecting their newsrooms to focus on the Web first, paper second . . . Their guiding principle: a constantly updated stream of intensely local, fresh Web content . . . Sites are fed with reader-generated content."[33] This style of journalism is sometimes called civic journalism, where the press participates actively in the public life of the geographical, political, and cultural communities it serves. The focus is to improve the communities and to give a voice to the residents. Pavlic explains that the New Media, however it is structured, "present an unprecedented opportunity for creating collaborative approaches to reporting."[34] The advent of mobile

[32] Holahan, "*Journal* Blazes," 11.
[33] Ahrens, "Newspaper Chain."
[34] Pavlic, *Journalism and New Media*, 104.

news reporting devices enable the creation of the "virtual newsroom" which allows

> journalists . . . to work entirely from the field without needing to enter a central newsroom location and to exchange messages, stories, and picture files with editors anchored firmly in cyberspace. [They] allow reporters in the field to have the same level of access to information, people, and processing power that central newsrooms have traditionally enjoyed . . . [which] enables breaking news stories to be managed from the field.[35]

Other forms of New Media are returning journalism to its grass roots. Citizen or community journalism (sometimes also called open source journalism) is taking root in many American towns and cities, where content is generated mostly by members of the community rather than by professional journalists. Some experts see this "as an advanced form of civic, public or communitarian journalism: involving the audience in the news, creating a kind of user-generated content."[36] Examples include iBattleborro.com, a citizen journalism site in Battleborro, Vermont, created in 2003, and the *Northwest Voice*, a community newspaper in Bakersfield, California.[37] Citizen journalism opens up the process, allowing people to create content by themselves for themselves, to celebrate their lives, to connect with their local community, and to discuss local political and social issues. J. D. Lasica, a founder of the grassroots media site Ourmedia.org, explains that citizen journalism is "giving people the power to contribute to the discussion in ways never thought possible before in traditional media models. . . . [w]ith the growing disaffection with media giants, the current generation of tools, and the unabated desire of people to tell their own stories, the movement should continue to grow and thrive."[38] Mary Lou Fulton, managing editor of the *Northwest Voice* in Bakersfield adds, "participation is something every newspaper should be doing. Through participation with readers, you get better, more relevant, and more authentic content than what you would have otherwise."[39] The wave of citizen journalism has led to the explosion of blogs on the Internet, as well as to personal ipod audio broadcasts, and eventually to sponsored Web sites such as Podshow and YouTube.

---

[35] Ibid., 106–7.
[36] Deuze, "Online Journalism."
[37] Miller, "Journalism Returns," 22.
[38] Quoted in Miller, "Journalism Returns," 27.
[39] Ibid.

The *Fort Myers News-Press*, with a circulation of 90,000, has tried several innovative civic journalism approaches. It created 14 full- and part-time mojos and is helping its 30 other reporters develop high tech expertise. It also enlists the help of dozens of reader experts (retired engineers, accountants, government insiders) to help review documents and conduct research for stories, a practice called "crowdsourcing." The *News-Press* and other Gannett papers are building searchable online databases on issues and topics to "enable people to do digging themselves and maybe find conclusions we won't. It's having thousands of investigative reporters instead of three," according to Michael Maness, Gannett's VP of strategic planning.[40] The *Valparaiso News*, in northwest Indiana, entered into a text messaging system with the city's police department. The department sends its own alerts, about road closings or accidents, for instance. Nevertheless, the newspaper controls its own alert system, using it to send out a text message about a commuter train hitting a freight train on a Sunday morning. The message built readership in the next day's print edition, according to the paper. The *Bakersfield Californian* gave each candidate in a local election a blog on the paper's Web site, and readers could ask candidates questions directly online.[41]

"Eager to lure new consumers, media outlets are experimenting with news-on-demand podcasts, [which] are basically digital files you can download from the Internet and listen to whenever your want."[42] Users employ free software to subscribe to regular podcast feeds, download them and transfer to a portable device like an iPod to listen to later. There were an estimated 5 million podcast users in the US in 2005, with predictions of 45 million by 2010. Not only media companies, but anyone with reasonable computer skills can produce one, and there are many such private producers. ABC News produces a weekly 15 minute podcast called the "ABC News Shuffle," as well as highlights from "Nightline' and "Good Morning America." CBS offers mostly features and commentaries, and NBC makes the "Nightly News" and "Meet the Press" available as podcasts, commercial-free. National Public Radio member stations offer almost 200 different podcasts. At least a dozen newspapers produce podcasts as well, including the Philadelphia Daily News. In addition to that, users can watch newscasts on their mobile phones by subscribing to provider services from the major networks. As Potter predicts, "when portable players go wireless and can download on the fly (which is already happening), 'news on demand'—both audio and video—will be a reality.

---

[40] Quoted in Ahrens, "Newspaper Chain."
[41] Frye, "Ever-evolving Newspaper."
[42] Potter, "iPod, You Pod," 64.

And the broadcast news organizations that buy into the idea now may ensure their own survival."[43]

The Future of New Journalism

With the development of new technologies and the spread of Internet usage, the future of New Journalism is unlimited. More and more American households have access to high-speed, high-powered Internet connections, and the continued development of portable devices make access to video, sound, and text as easy as watching TV or reading a magazine or newspaper. As Roberts explains, "news broadcasters lately are seeing the future. And guess what? It's not in the evening—and maybe not on the TV screen. Instead, broadcast journalism is going multiplatform. From ipods and cell phones to laptops, NBC, ABC and CBS are digitalizing their high-priced anchors . . . for news-on-the-go."[44] CBS News is planning to launch an online news broadcasting service using broadband technology to revitalize its news operations. According to Greppi, "CBS News, the TV news organization perhaps most defined by its past, is going online in a big broadband way in an attempt to redefine, revitalize and perhaps even reinvent itself as a 24-hour high-speed news organization."[45] Content will include Web logs by reporters, a blog called "Public Eye" to facilitate dialog with the public about CBS News, and a "video jukebox" that will enable users to build their own personal newscast.[46] The service will be supported by advertising and be free to users. ABC already has a broadband service, ABC News Now, that is available to more than 30 million online and mobile subscribers. NBC has put its entire evening newscast online and Brian Williams podcasts and blogs.

News organizations must be willing to adapt and to understand the characteristics of Internet news users and distribute information in a variety of formats. According to Roger Fidler, director of technology initiatives at the Missouri School of Journalism, "the reality is that for the future, there won't be any single product that will capture a large audience. To survive, it means that newspapers (and broadcasters) need to be able to use a number of devices."[47] Palser recommends that news sites make better use of material provided by YouTube, Flickr, Google Maps, and

---

[43] Ibid., 64.
[44] Roberts, "Future of Evening News," 44.
[45] Greppi, "CBS News Plans," 6.
[46] Ibid., 24.
[47] Quoted in Frye, "Ever-evolving Newspaper."

other Internet sites in a process called "mashing." The idea is for news sites to "mash" their own content with Internet sites to, for instance, create geographical guides to restaurants or events, or present the location of an occurrence. She explains that news organizations need to abandon their old instinct to control and own everything that they do and should instead take advantage of open-source technology and personal media sites. "Rather than expecting people to submit their content to [them], news organizations need to start going where the people are," she says.[48]

Problems and Opportunities

The state of the news business in the United States has been problematic for a long time. Both newspaper readership and network news broadcast audiences have declined significantly during the past 25 years. Academics and journalists have rued the decline in the quality of news as well, pointing out that audience fragmentation, trends toward entertainment, a focus on profit and advertising have dummied down content and lowered news standards. Downie Jr. and Kaiser say that economic factors have held sway over American journalism for some time, bringing changes to the industry that have too often been for the worse: smaller news staffs, less ambitious coverage, and a change in focus, where the public "need to know" is replaced by the "want to know." This has led, they argue, to a crisis in confidence and a fear that the industry will lose all credibility. Another concern is that the Internet and New Media can further exacerbate the situation, rather than improve it.

The sheer volume of information on the Internet, much of it untraceable, coupled with the trend toward citizen journalism, where everyone is a journalist, has led to worries about the quality of online news. Pavlic makes the point that "much Internet content is of dubious or unknown origin, sometimes worthless or intentionally misleading."[49] Gup worries that "if online users are unable to discriminate between the credible and misleading, the public could tune us all out, damaging public discourse and producing not an informed citizenry but an audience of cynics."[50] Journalists often feel that the virtues of restraint, of accuracy, of confirmation, and of balance and fairness can be lost in such an environment. They wonder if the suppliers of information on the Internet are worthy of public trust and aware of the special responsibilities that

[48] Palser, "Missed Opportunities," 84.
[49] Pavlic, *Journalism and New Media*, 63.
[50] Gup, "Who's a Journalist?—I," 42.

accompany it. They use the example of Matt Drudge and similar chat room gossipers and bloggers whose exclusive stories sometimes contain a grain of truth and other times are wildly wrong. "Like reporters, he disseminates what he finds out, but he does not necessarily take pains to verify his items, and he operates without an editor or other gatekeeper,"[51] explains Tom Goldstein, dean of Columbia University's Graduate School of Journalism. Many journalists oppose the use of open sources, arguing that it is an amateur journalism produced by citizens, activists, and troublemakers with personal agendas. Others are upset about such practices as mojos' posting stories directly to the Web without any editing, a common practice at some newspapers. Pavlic adds that journalists need to verify online information from off-line sources and should never rely exclusively on online information for a story. Digitalization adds to the problem, making it incredibly easy to fabricate photographs, video and audio recordings, creating artificial realities to the point where you cannot be sure of anything that you see or hear on the Internet.

Other professionals worry that the focus on ads, profit, and attracting an audience fits neatly into the Internet culture and contributes to the decline in news reporting rather than countering it. The constant updates and feeds, the repackaging of stories for different readers and different viewer usage patterns is really designed to keep consumers viewing ads, according to many critics. It is all about making money for the stockholders, not creating responsible news reporting. The up-to-the-minute news cycle has too often resulted in "news [being] displaced by hyperbole, rumor, and innuendo as if the technology had caused a cultural stroke," adds Carey.[52] There is a general worry that the zeal to feed the Web with fresh material and updates contributes to the publishing of "fluff" in addition to news, adding to the increase in feature, consumer product, and celebrity reporting at the expense of hard news.

What is emerging, according to Pope, is a made-to-order news landscape "spun out of the Internet and digital cable television. News, like much of the rest of pop culture in America, is becoming a niche game."[53] These niche products, aimed directly at specifically targeted audiences tend to undermine traditional, broadly targeted news products. The fragmenting of the news audience, the impact of advertising, and the added competitive dimension of providing continual reporting has led to a relentless competition in the news business and to what many critics see as a lowering of standards and a change in the definition of news. Easy

---

[51] Quoted in Downie Jr. and Kaiser, *News About the News*, 206.
[52] Carey, "Lawyers, Voyeurs and Vigilantes," 23.
[53] Pope, "Network and Cable TV," 64.

access often means shoddy news, where anything goes. Multiplatform reporting can also lead to problems when the focus oftentimes is less on content and more on connectivity, making a news site "hyperadaptive" with high-tech interactivity, hypertextuality, and multimediality to appeal to a wide range of viewing patterns. Says Brian Storm, director or multimedia at the MSNBC.com newsroom, "we get distracted from the storytelling by the technology, its possibilities, limitations, and decisions" about how to us it.[54]

The Anna Nicole Smith story can be used as an example of the Internet's influence on current American journalistic practices. An article in the February 10, 2007 edition of the *Los Angeles Times* explains how Internet traffic can influence front-page news. Human interest and curiosity have always been ingredients of American journalism, but what was different in this case, the author maintains, is the way in which the Smith story made the jump from tabloid covers to the pages of serious newspapers, as well as the airwaves of evening news reports. "The mainstream journalistic coverage of Smith's death is among the first such stories driven, in large part, by editorial perception of public interest derived mainly from the Internet."[55] What happened, Rutten goes on to explain, is that news editors across the USA watched the number of "hits" recorded on Internet sites covering the Smith story, and those numbers helped shove the story onto the front pages of major newspapers. "What makes this of more than passing interest is that serious American journalism is in the process of transforming itself into a new hybrid news medium that combines traditional print and broadcast with a more purposefully articulated online presence," the article continues. The ability of the Web to gauge readers' interest in a particular story on a minute-by-minute basis leads to ratings game of sorts, Rutten explains, which can be used as marketing information by news organizations. Such preoccupation with ratings, the focus on building an audience and attracting advertising, is affecting what is being reported by America's mainstream news organizations. "It's a good moment for American newspapers to take a reflective breath to consider just how they want to play this numbers game—or, more important, whether they want to play it at all," concludes Rutten.[56]

Another *Los Angeles Times* article by media professor and author Neal Gabler concerns the real power of the Internet to affect culture and even to create its own culture. He explains the effect of New Media on the

---

[54] Quoted in Downie Jr. and Kaiser, *News About the News*, 199.
[55] Rutten, "How Smith's Death," E12.
[56] Ibid., E12.

American movie industry, arguing that something has fundamentally changed in our relationship with the movies. According to Gabler, we do not discuss movies the way we once did. Films used to be America's "Democratic Art," displaying its character and national consciousness, situated at the cultural center of the country. Now they increasingly sit at the margins. Now it is more important and more gratifying, to young people especially, to know which celebrities are dating each other, who is in rehab or who has had plastic surgery. Now, "the intrinsic value of a movie, or of most conventional entertainments, has diminished. Their job now is essentially to provide stars for *People*, *Us*, 'Entertainment Tonight' and the supermarket tabloids, which exhibit the new 'movies'—the star's life sagas."[57] In other words, the movies generate publicity for the tabloids and entertainment shows, rather than the other way around.

Movies once were the art of the middle class, providing a common experience and language, a sense of unity. Now people identify themselves as members of smaller and smaller cohorts, part of the fragmentation of audiences and of American society, according to Gabler, part of the impulse toward promoting the individual rather than the community. Furthermore, he says, "there is now an instrument to take advantage of the social stratifications. To the extent that the Internet is a niche machine, dividing its users into tiny, self-defined categories, it is providing a challenge to the movies that not even television did, because the Internet addresses a change in consciousness."[58] The Internet not only creates niche communities, it plays to another power force in modern America: narcissism and celebrity culture. Web sites such as Facebook, MySpace, and YouTube, where the user is the star and content is democratized, get far more hits than movies get audiences. These sites are replacing one of the movie's main functions. "If stars' lives are superseding movie narratives, audiences are superseding the stars. In effect, we have become our own movies."[59] This is the power of the Internet, according to Gabler. It is a change in consciousness, not just a change in technology.

One more significant change in journalistic practices caused by Internet technology has been identified in a *Washington Post* article. "The humble interview, the linchpin of journalism for centuries, is under assault," it says.[60] In the digital age, the balance of power has shifted from the person asking the questions (the journalist), to the person being interviewed and to the readers. Feeling that they may be misquoted or

---

[57] Gabler, "Hollywood's Fade to Black," M4.
[58] Ibid.
[59] Ibid.
[60] Kurtz, "Interviews, Going the Way," C1.

misrepresented in a published interview, some celebrities, business executives, and politicians are saying that they will only respond to email questions. This gives them the possibility of posting the complete interview online if they feel that they need to respond to or correct an impression made by the published version. While some critics rue this as a further diminishing of the role of professional journalists, others hail it as a step toward increased accuracy and access. According to Jay Rosen, a journalism professor at New York University, "readers have more power because they have more sources, and sources have more power because they can go direct to readers."[61]

So, is the Internet a negative power in American culture, creating problems of validity, undermining standard journalistic practices, and contributing to the fragmentation of the society and the emphasis of profit over responsible reporting? Or is it the ultimate democratizing force that opens up journalistic practices, creates unlimited access to information, and provides quality, contextualized, interactive news coverage? Part of the answer lies with the major media companies. News organizations need to play the role of watchdog over Internet content, helping users determine what to trust. They also need to provide exclusive, significant content and multimedia coverage that can lure readers and advertisers away from blogs and chat rooms. Traditional practices of accuracy, fairness, and balance need to be emphasized. "The more raw information there is available, the more consumers will need professional journalists to sort through it for them . . . organize the important information in an easily digestible form, check to be sure it's accurate, and display it in a way that reflects its importance."[62] It is arguable that the Internet helps more than it hinders, that it provides a richer news diet, that it is much easier to be informed today than it was 30 years ago. The high profitability of the best news organizations shows that quality news can make a profit. However, owners need to realize that high-quality journalism attracts readers and advertisers. Part of the answer lies with the American people. Will Americans want to take advantage of the huge capabilities of the Internet for higher-quality news coverage? Or will they prefer to use the Internet primarily for entertainment, the way they use television?

---

[61] Quoted in Kurtz, "Interviews, Going the Way," C3.

[62] Downie Jr. and Kaiser, *News About the News*, 256.

Works Cited

Ahrens, Frank. "A Newspaper Chain Sees Its Future, and It's Online and Hyper-Local." *Washington Post*, December 4, 2006.

Carey, James W. "Lawyers, Voyeurs and Vigilantes." In Giles and Snyder, *What's Next*, 19–28.

Christopher, L. Carol. "Journalists Learn to Become Storytellers." *Seybold Report*, January 11, 2007.

Deuze, Mark. "Online Journalism: Modelling the First Generation of News Media on the World Wide Web." *First Monday* 6, no. 10 (October 1, 2001). http://firstmonday.org/htbin/cgiwrap/bin/ojs/index .php/fm/article/view/893/802.

Downie, Leonard, Jr., and Robert G. Kaiser. *The News About the News*. New York: Knopf, 2002.

Flavian, Carlos, and Raquel Gurrea. "The Choice of Digital Newspapers: Influence of Reader Goals and User Experience." *Emerald* 16, no. 3 (2006): 231–47.

Frye, Michele Holtkamp. "The Ever-evolving Newspaper." *Quill*, December 1, 2006.

Gabler, Neal. "Hollywood's Fade to Black." *Los Angeles Times*, February 25, 2007.

Giles, Robert, and Robert W. Snyder, eds. *What's Next? Problems & Prospects of Journalism*. New Brunswick, NJ: Transaction Publishers, 2001.

Greppi, Michele. "CBS News Plans Major Online Push." *TelevisionWeek*, July 18, 2005.

Gup, Ted. "Who's a Journalist?—I." In Giles and Snyder, *What's Next*, 39–44.

Holahan, Catherine. "*Journal* Blazes Newspapers' New Trail." *BusinessWeek*, January 3, 2007. http://www.businessweek.com /technology/content/jan2007/tc20070102_893652.htm.

Kansas, Dave, and Todd Gitlin. "What's the Rush?" In Giles and Snyder, *What's Next*, 83–90.

Katz, Jon. "The Future is the Net." In Giles and Snyder, *What's Next*, 15–18.

Kebbel, Gary. "The Challenge of Community Building." *Nieman Reports* 60, no. 4 (2006): 27–28.

Kurtz, Howard. "Interviews, Going the Way of the Linotype?" *Washington Post*, May 21, 2007.

Miller, Ron. "Journalism Returns to Its (GRASS) Roots." *Econtent* 28, no. 6 (2005): 22–27.

Naylor, Richard, Stephen Driver, and James Cornford. "The BBC Goes Online: Public Service Broadcasting in the New Media Age." In *Web.Studies: Rewriting Media Studies for the Digital Age*, edited by David Gauntlett. London: Arnold, 2000.

Newmark, Craig. "Community Building on the Web: Implications for Journalism." *Nieman Reports* 60, no. 4 (2006): 25–27.

Palser, Barb. "Missed Opportunities." *American Journalism Review* 28, no. 4:84.

Pavlic, John V. *Journalism and New Media*. New York: Columbia University Press, 2001.

Pew Research Center for the People & the Press. "Internet News Audience Highly Critical of News Organizations." Pew Research Center Publications, August 9, 2007. http://pewresearch.org/pubs/564/Internet-news-audience.

Pope, Kyle. "Network and Cable TV." In Giles and Snyder, *What's Next*, 63–70.

Potter, Deborah. "iPod, You Pod, We All Pod." *American Journalism Review* 28, no. 1:64.

Roberts, Johnnie L. "The Future of Evening News." *Newsweek*, March 17, 2006.

Rutten, Tim. "How Smith's Death Hit Page 1." *Los Angeles Times*, February 10, 2007.

Stevens, Jane Ellen. "Taking the Big Gulp." *Nieman Reports* 60, no. 6 (2006): 66–69.

Fan Culture in the Internet

Ewa Grzeszczyk

Fan cultures, previously only peripheral in academic inquiry, in recent years have been the subject of serious consideration not only by scholars but also by journalists, lawyers and mass culture producers.[1] As Henry Jenkins claims, such increased attention is to a large extent the result of the movement of much fan activity into the Internet. The virtual realm has not only granted fans great possibilities of creation, self-expression, communication and activism but also considerably increased their visibility.[2]

In my essay I will focus on the changes fan communities have been experiencing since they started to make use of the potential the Internet offers. I will discuss the ways in which the Net has transformed fandom and analyze selected forms of fan activity online, such as fan fiction, hackfiction, spoiling and fan activism against certain decisions of media corporations. I will discuss issues such as collective intelligence and subcultural capital and point at the hierarchies which exist within fan communities. The reactions of media producers towards the increasing significance of online fandom will be also examined. Having analyzed new forms of cultural production of fans active online, I will pose a question whether the majority of media consumers can be described as an active audience.

Fans Online

The growing popularity of the Internet has dramatically changed the experience of being a fan. Nowadays fans from different geographic locations can discuss their favorite media texts with a large group of like-minded enthusiasts not only at fan conventions organized a few times a year but every time of night and day.[3] Writers of fan fiction can post their

---

[1] Jenkins, *Convergence Culture*, 246.
[2] Jenkins, *Convergence Culture*; Jenkins, *Fans, Bloggers and Gamers*, 140–44.
[3] Jenkins, *Fans, Bloggers and Gamers*, 142.

works online within minutes and in some more minutes receive audience feedback. Fans can create their own Web pages or use the already existing ones, where various information about their beloved popular culture texts can be easily found. As Suzy Shaw noticed already in 1994, "What once might seem like isolated obsession now can be shared with people who not only understand, but actually encourage and reinforce the enthusiasm."[4]

No longer isolated individuals, contemporary fans can create what Pierre Lévy calls "knowledge communities" which are based on common intellectual investment and emotional engagement.[5] Such communities are voluntary, temporary and task oriented. Members are emotionally invested but can belong to many knowledge communities at the same time and also shift from one to another as their interests change.[6] Members of knowledge communities can make use of "collective intelligence" which for Lévy stands for the collective expertise of virtual communities: no-one knows everything but each of us knows something and a sum total of this knowledge can be accessed in response to a specific question.[7] According to Nancy Baym, fans make their expertise broadly available because they are motivated by "epistemaphilia"—the pleasure in sharing knowledge.[8] Moreover, the Internet as a channel of quick and effective communication has enabled fans to engage in what Matthew Hills names "just-in-time fandom": for example the enthusiasts of a certain TV show can comment on and discuss their favorite program while it is being aired or immediately afterwards.[9]

Transformation of fan production and its distribution are still other results of fan communities' engagement with the Net: fanfiction online nearly completely replaced that printed in fan zines; fan-generated music (filk), movies or photo collages can be quite easily downloaded from relevant Web sites.[10]

As the Internet fandom blooms and diversifies, it is no longer an activity at the margins of culture but enters the mainstream.[11]

---

[4] Shaw, "Internet as an Entertainment System," 9.

[5] Jenkins, *Convergence Culture*, 27; Jenkins, *Fans, Bloggers and Gamers*, 137. Henry Jenkins refers to Lévy's book *Collective Intelligence*.

[6] Ibid.

[7] Jenkins, *Convergence Culture*, 4, 26–27; Jenkins, *Fans, Bloggers and Gamers*, 139.

[8] Jenkins, *Fans, Bloggers and Gamers*, 139, referring to Baym, "Talking about Soaps."

[9] Hills, *Fan Cultures*, 141–42; Jenkins, *Fans, Bloggers and Gamers*, 118, 141.

[10] Jenkins, *Fans, Bloggers and Gamers*, 143.

[11] Ibid., 142.

Contesting the Dominant Reading? Fan Fiction

The term "fan fiction" generally refers to reader-generated stories based on characters from TV, books and movies.[12] Its roots can be traced back to the photocopied stories inspired by *Star Trek*, passed around at fan conventions or via snail mail in the 1970s.[13] In recent years fan fiction (in short "fanfic") has enjoyed considerable media attention, largely due to the growing popularity of the form in the Internet.

Fan fiction writers can be inspired nearly by anything from the Bible to manga. Ideas for writing might come from classic novels and science fiction, TV shows such as "Friends" and computer games such as "King of Fighters." There are also Song-fics: stories inspired by lyrics and real person fiction: here Princess Diana, Bill Clinton and the Beatles have been powerful sources of inspiration.[14] As "[i]t is universally acknowledged that Jane Austin wrote too little,"[15] her fans, known as Janeites can publish their Austen-inspired stories at Austin.com or The Republic of Pemberley (www.pemberley.com).[16] Recently one of the most stimulating sources is the Harry Potter series. Potter fanfic can be found at a number of sites, for example MuggleNet.com, harrypotterfanfiction.com or The Sugar Quill.

Fan fiction can have a number of forms. One of the most popular is a Mary Sue narrative. Lieutenant Mary Sue became the commander of the Starship Enterprise in a story written in 1974 by a trekkie—that is, a *Star Trek* fan—Paula Smith.[17] Today "Mary Sue" stands for an idealized version of the author featuring prominently in the narration or simply, for narcissistic fan fiction.[18] Mary Sues are often employed in a fanfic format which is based on retelling the original story from a different character's point of view. Other fan fiction forms include prequels (back stories of central characters) and sequels, or they provide the reader with "missing scenes."[19] Popular are stories created by "shippers" imagining a relationship between characters.[20] One sub-genre of this type, called "slash," depicts same-sex relationships between protagonists, for example

---

[12] Young, "Fan Fiction Phenomena."
[13] Jenkins, *Fans, Bloggers and Gamers*, 42.
[14] Pugh, "Democratic Genre."
[15] O'Connell, "World Without End."
[16] Ibid.
[17] Chander and Sunder, "Everyone's a Superhero," 598.
[18] Ibid., 597, 599.
[19] O'Connell, "World Without End."
[20] Short for "relationship" or "relationshippers." Jurgensen, "Rewriting the Rules of Fiction"; Scodari and Felder, "Creating a Pocket Universe."

Kirk/Spock (K/S) or Potter/Malfoy and is most often written and read by young women.[21] Another sub-genre is referred to as hurt/comfort (h/c): in this kind of fan fiction usually a male hero suffers severe physical/emotional (or both) damage and then is being rescued and consoled by another character.[22] The fantasy of fanfic writers has no limits; in the so-called "cross-overs" characters from different texts can interact in one story, for example in one narrative Buffy the Vampire Slayer meets Mr. Darcy.[23]

Most fanfic is created by women, and unfortunately, much of it is of low quality. At the popular fanfic sites one may publish his or her story in a matter of minutes, and readers sometimes congratulate the author on good grammar.[24] Some sites offer writers the assistance of the so-called "beta readers" and are selective in their choices of what to archive.[25] Henry Jenkins in his analysis of *Harry Potter* fanfic, composed mainly by children and teenagers, points to the positive aspects of this kind of creativity: acquiring writing and editing skills and learning to voice one's concerns.[26]

Writers have presented different attitudes to fan fiction.[27] Joan Rowling has been sympathetic to the efforts of fan writers. Another popular writer, Ann Rice, however, has clearly indicated that she does not tolerate fan fiction and advised amateur writers to create their own stories. At her Web page one can find the following statement: "I do not allow fan fiction. The characters are copyrighted. It upsets me terribly to even think about fan fiction with my characters. I advise my readers to write your own original stories with your own characters. It is absolutely essential that you respect my wishes."[28] A fantasy novelist Chelsea Quinn Yarbro who believes that fanfic is a form of robbery assumes that her lawyer has sent about 20 "cease and desist" letters prompting some fanfic Web pages and authors to take their writing down.[29]

---

[21] Walker, "Harry Potter Comes of Age."
[22] Pugh, "Democratic Genre," 6.
[23] Ria, "Lady for a Day-dream," *Twisting the Hellmouth*, September 21, 2007, http://www.tthfanfic.com/Story-14021/Ria+Lady+for+a+Day-dream.htm (accessed March 24, 2009).
[24] Young, "Fan Fiction Phenomena."
[25] Ibid.
[26] Jenkins, *Convergence Culture*, 169–205.
[27] Walker, "Harry Potter Comes of Age"; Jurgensen, "Rewriting the Rules of Fiction"; Young, "Fan Fiction Phenomena."
[28] Ann Rice's official Web site.
[29] Jurgensen, "Rewriting the Rules of Fiction."

Much of fanfic analysis concentrates on the ways such kind of creativity enables marginal audiences to create "alternate universes" and in this manner challenge the hegemonic reading. After all, as Jenkins observed: "Fandom . . . is born of a balance between fascination and frustration . . . ."[30]

Approaches focusing on subversive readings are especially common in the American Cultural Studies audience research tradition associated with Fiske[31] and were employed by such scholars as Jenkins in his seminal work on fandom *Textual Poachers* or Brown in *Soap Opera and Women's Talk: The Pleasure of Resistance*.[32]

The presumption of this kind of analyses is that a media text, because of its polysemic nature, allows for multiple readings so fans are empowered as they have autonomy in constructing the meaning.[33] For example, Chander and Sunder in their analysis of Mary Sues assume that such stories, although often ridiculed, are not only exercises in narcissism. Writing Mary Sues gives the authors the ability of reworking the canon to valorize women and marginalized communities.[34] As another author notes, fans are often "more in tune with the zeitgeist" than for example TV producers. For example, long before the 1990s a big part of the audience wanted stronger female characters and more vulnerable males: what they could not find on TV they created in their fanfic.[35] Still, some authors, like McKinley in her study of *Beverly Hills 90210*, have shown that fan readings may differ from the dominant reading only very little.[36] Moreover, some authors doing research on fanfic observe that its readers do not tolerate well fiction which drifts too far away from the institutionally produced original. For example, says one *X-Files* fan "I'm finding it pretty hard right now to read fanfic in which M&S are 'together,' because the show isn't progressing in that direction and I like relatively realistic fanfic."[37]

Very often the readings generated by fans are reflective of the fan's self. For example Sandvoss observed that fans of Chelsea football club described the club accordingly to their own sense of the self. The fan who referred to the club as successful was a successful businesswoman herself,

---

[30] Jenkins, *Convergence Culture*, 247.

[31] Scodari and Felder, "Creating a Pocket Universe," 238.

[32] Ibid., see Jenkins, *Textual Poachers* and Brown, *Soap Opera and Women's Talk*.

[33] Sandvoss, "One Dimentional Fan," 283–84.

[34] Chander and Sunder, "Everyone's a Superhero," 597–611.

[35] Pugh, "Democratic Genre."

[36] McKinley, *Beverly Hills*, quoted in Sandvoss, "One Dimentional Fan," 824.

[37] Scodari and Felder, "Creating a Pocket Universe," 249.

whereas the one who stressed the club's unfulfilled potential was a not-that-financially-successful freelance writer. A fan of Mediterranean descent focused on the cosmopolitan character of the club, while for another the club was an extension of his own belief in the English supremacy.[38] In short, these different readings "tell us less about the club than the individual fan."[39] For Sandvoss, different interpretations of a single fan text or object "are reflective of fan's sense of self and corresponding horizon of experience."[40] Moreover, as he notes, fans often look for parallels between themselves and characters from films or TV shows. A similar mechanism was observed in the context of interpreting song lyrics. As one author explains, "Fans always have the feeling that Bruce [Springsteen] is reading their minds because 'their minds' are active elements in constructing and interpreting the music."[41]

Subcultural Capital—Hack Fiction and Spoilers

Apart from the subversive reading approach, another significant trend in studying online fandom is connected with the concept of subcultural capital.[42] Drawing on the notion of Bourdieu's "cultural capital," Sarah Thornton coined the term "subcultural capital" which refers to status within relevant subculture because of one's being "in the know," appropriate looks and behavior.[43] Media consumption is crucial for the creation of this kind of capital as "the difference . . . between being in and out of fashion, high or low in subcultural capital, correlates in complex ways with degrees of media coverage, creation and exposure."[44] As Thomas has observed, "'subcultural capital' . . . is also useful in the discussion of fan cultures generally, since it implies an assertion of difference and status which would not be recognized by the wider society, given the generally low prestige of fan cultures."[45] According to Jenkins,

---

[38] Sandvoss, "One Dimentional Fan," 832.

[39] Ibid.

[40] Ibid.

[41] Cavicchi, *Tramps Like Us*, 135, quoted in Sandvoss, "One Dimentional Fan," 832–33.

[42] For a broader discussion of the use of the notion of subcultural capital in studies of fan cultures, see Williams "'It's About Power.'"

[43] Thornton, "Social Logic," 186–87.

[44] Ibid, 187.

[45] Thomas, *Fans, Feminisms*, 10, quoted in Williams, "'It's About Power.'"

fans have been traditionally a marginalized group, perceived as "scandalous category" advised to "get a life."[46]

As Williams has observed, fan cultural and discursive power comes from their knowledge about the object of their fandom and their ability to control the discourse of other fans.[47] At the top rungs of the fan hierarchy we can find "executive fans" or "net celebrities" such as fans who run their own Web sites, fan activism campaigns leaders, spoiler sources and hackers. According to Williams, executive fans inspire and control fan community discourse: they exercise power over other fans by control of information (subcultural capital) and social constraints (social capital).[48] Executive fans establish the rules of netiquette; they may ban the discussion of certain topics, reveal as much information as they want and at the pace they choose, they can also shun or ridicule newbies or fans proposing a reading they consider inappropriate. Fans usually try to avoid conflicts with "net celebrities" as they may negatively influence their social position among other fans: banned from one site, a fan can be also ostracized at a number of related sites.[49]

The two groups of fans that can well illustrate the notion of subcultural capital are fan fiction hackers and spoiler sources.

As Fiona Carruthers explains, fanfic hackers search the net looking for examples of "bad" fan fiction and insert their mocking comments into the body of the hacked text.[50] Being a fanfic hacker requires the possession of subcultural capital on at least two levels: the knowledge of the pop culture text which was the inspiration for the fan fiction author and the conventions of the fanfic genre. Hackers criticizing what they label as "suckfic" possess the competence of expression and the conviction that they occupy a high position in the hierarchy of fanfic consumers. However, sometimes they lament the loss of time they could devote to reading/writing better fan fiction. Fan fiction hackers struggle for distinction: high position within the hierarchy of other fans, and who is or is not a legitimate fanfic writer. That is why "Hackers punish the incompetence of other fans because they pose a threat to what it means to be intelligent 'readers' of the text."[51] Hackers act as a kind of gatekeepers, marginalizing the readings they find inappropriate. Carruthers is of the opinion that for hackers "fan fiction . . . is justifiable as long as it

[46] Jenkins, *Textual Poachers*, 15, as observed by Williams in "'It's About Power.'"
[47] Williams, "'It's About Power.'"
[48] Ibid.
[49] Ibid.
[50] Carruthers, "Fanfic is Good."
[51] Ibid.

competently represents the author's voice and adds a new textual meaning to the show itself."[52] Still, she quotes the opinion of one hacker who says "Hacking fan fiction is like a therapy. It lets you get all the bad feelings that you are left with after a horrible fic that violates one of your favorite TV shows or books out of your system."[53]

Spoiler sources are another example of fans possessing a degree of subcultural capital which enables them to occupy a significant position within the fan community. Spoiling is an activity connected especially with TV fandom online: spoiler sources deliver pieces of information (called "spoilers") on upcoming development in the next episodes of the show revealed to fans before a given episode is actually aired.[54] The sources often provide an outline of what is going to happen, and the pleasure for the fans is to evaluate the evidence and speculate about the details.[55] Spoilers have to be verified to be recognized as reliable,[56] especially because from time to time spoiler hoaxes crop up—misleading clues placed either by network executives or other fans.[57]

Henry Jenkins in his *Convergence Culture* describes the spoiling community of *Survivor* enthusiasts. Hard-core fans try to find out the location of the show's production, the names of the contestants, the sequence in which they will be eliminated, the final four contestants and finally, the winner, before the producers reveal it. Spoiler sources use various tactics and sources of information: satellite photographs, show's frame by frame analysis (to look for hidden clues), leaks from the show's insiders and hacking.[58] *Buffy the Vampire Slayer* spoilers analyzed by Williams could concern episode titles, plot outlines, characters' reappearances or deaths and romantic developments. Certain fans admitted reading spoilers can be addictive.[59]

Both authors in their research on spoiling communities notice a distinct hierarchy of power and prestige existing within the group. At the top there are the best informed sources (they sometimes work for the industry producing the show) and those enjoying biggest respect of other fans because of their former spoiling achievement.[60] Sometimes the top-rank

---

[52] Ibid.

[53] Nerbie, quoted in Carruthers, "Fanfic is Good."

[54] Williams, "'It's About Power'"; Jenkins, *Convergence Culture*, 30.

[55] Williams, "'It's About Power.'"

[56] Ibid.

[57] Jenkins, *Convergence Culture*, 46–49.

[58] Ibid., 25–58.

[59] Williams, "'It's About Power.'"

[60] Williams, "'It's About Power'"; Jenkins, *Convergence Culture*, 46–49.

spoiler sources may form a clique like in case of the so-called "brain trusts" at www.survivorsucks, where the smartest and most articulate do the most hard-core investigation and communicate with each other using password-protected sites.[61] Lower in the hierarchy are the consumers of the information—spoiled fans (who sometimes, especially if addicted to the news, refer to themselves as "spoiler whores") and at the bottom, the "innocent," unspoiled fans.[62]

Fan Advocacy

The visibility of fan communities and their practices online made consumer activism more effective. I will discuss two examples of fan advocacy which was facilitated by the Internet: TV fans' advocacy to prevent the cancellation of their favorite show and teenage fanfic writers' reaction to the attempts of big media companies to control their work.

According to Scardaville, most often TV primetime fans organize big scale around the issue of show cancellation—usually their activism begins before the show is cancelled, fans can learn about it from leaks from people professionally tied to the show or media reports.[63] In her essay "Accidental Activists" the author describes the attempts of fans to prevent the cancellation of the *Another World* TV show (most of her respondents were devoted fans who have been watching the show for over 20 years). After cancellation of the show fans advocated for bringing it back to the air and for finding new jobs for certain actors. Groups of fans maintained their Web pages (and usually a listserv) informing about current activities. No money was being collected. Sympathetic insiders, such as actors, crew members or the network employees, could provide the fans with current information, advice and instruct them on how to conduct a certain action, for instance how to write a protest letter. Most of the information was disseminated online but most of the action took place offline (petitions, calls to the network, talking to reporters, wearing special ribbons, and a consumer boycott of particular network products). Eventually, the show was re-aired on an affiliated soap-opera network.[64]

Fans try to influence TV producers, making their watching preferences visible online. In 2004 the Web site savethatshow.com introduced at fans' demand a new feature where fans could vote to save a show.[65]

---

[61] Jenkins, *Convergence Culture*, 38–40.
[62] Williams, "'It's About Power.'"
[63] Scardaville, "Accidental Activists," 887.
[64] Scardaville, "Accidental Activists."
[65] Ibid, 887.

Fan advocacy emerges also among the youngest fans. As Jenkins observes in his analysis of Harry Potter fandom, most of the book series fans are children and teenagers and although J. K. Rowling presents a welcoming policy towards Potter fanfic, the authors faced serious problems when Warner Bros. acquired the film rights in 2001.[66]

Heather Lawver, an American teenage creator of *The Daily Prophet* (an online magazine where children and teens contribute their essays), organized The Defense against Dark Arts organization to protect fans threatened with legal action.[67] She stated:

> Warner was very clever about who they attacked. . . . They attacked a whole bunch of kids in Poland. How much of a risk is that? They went after the 12 and 15 years olds with their rinky-dink sites. They underestimated how interconnected our fandom was. They underestimated that we knew those kids in Poland and we knew the rinky-dink sites and cared about them.[68]

As it transpires from the above example, fan solidarity and activism thanks to the Internet can develop even on the intercontinental scale.

Signs Indicating that Media Producers Take into Account the Voices of Fans

Media producers and advertisers appreciate the potential of fan culture; in their attempts to build a long-term relation between a consumer and a brand they increasingly realize that the most loyal consumer is a fan.[69]

Popular culture producers and marketers more and more often employ strategies such as "permission based marketing," "relationship marketing" and "viral marketing" based on the concept of engaging consumers into the production and marketing of media content.[70] Such strategy is especially important when we take into consideration that fans have traditionally been early adopters of new media technologies and also most creative part of the audience.[71]

A good example of fan culture incorporation into the mainstream of cultural production is the case of fanfiction. Although professional writers

---

[66] Jenkins, *Convergence Culture*, 185.
[67] Ibid., 185–86.
[68] Quoted in Jenkins, *Convergence Culture*, 186.
[69] Jenkins, *Convergence Culture*, 63.
[70] Jenkins, *Fans, Bloggers and Gamers*, 147–48.
[71] Jenkins, *Convergence Culture*, 12, 131; Jenkins, *Fans, Bloggers and Gamers*, 138.

are being advised not to read fanfic as they could be accused of profiting from other people's concepts,[72] it has been noted that TV show runners have been long fishing for ideas at fan fiction sites.[73] Many media companies, usually eagerly chasing those who might make an unauthorized use of their copyrighted material are adopting towards fan fiction an attitude of "benign neglect," as they are aware of the strong fan base such authors represent.[74] Moreover, several fanfic writers signed high-paying contracts for their original works or fiction inspired by older classics no longer protected by copyright law,[75] for example Pamela Morgen, a librarian from Idaho who after posting for 10 years stories about one *Pride and Prejudice* character was offered $150,000 to publish them in the form of three novel series.[76]

Sometimes media producers take active steps to exploit fanfic authors' creativity. This happened in case of FanLib, a company working together with networks and producers and organizing writing contests for fanfic authors. When the company announced a contest for writing a scene for "The L Word" TV drama in 2006, as many as 20,000 people registered.[77] Network executives also expected that FanLib, targeted mainly to young people, ages 13 to 20, can boost the ratings of newer shows, especially those aimed at younger audience.[78] In short, FanLib was to be a great marketing tool. However, the fans have discovered that the company wanted to cynically capitalize on their creativity and actively inform the fanfic community about it.[79]

Not only the potential authors of TV shows' scripts but TV fans in general are an excellent illustration of the complex relations between media executives and the fan community. Interviews conducted by Scardaville with people who have worked in the soap opera industry indicate that the Internet has significantly changed soap fans' influence on producers: if in the 1970s and 1980s fans were in no position to change story lines, in recent years the industry executives pay more attention to fan opinions, as "it's like having a focus group right at your fingertips on

---

[72] Jurgensen, "Rewriting the Rules of Fiction."
[73] Friedman, "Fan Scripts."
[74] Jurgensen, "Rewriting the Rules of Fiction."
[75] Young, "Fan Fiction Phenomena."
[76] Jurgensen, "Rewriting the Rules of Fiction."
[77] Ibid.
[78] Friedman, "Fan Scripts."
[79] See for example icarusancalion, "Article Summing up FanLib," *Live Journal*, comment posted May 20, 2007, http://icarusancalion.livejournal.com/626928.html (accessed March 24, 2009).

the Internet."[80] Fans have been noted to influence the storyline and the fate of shows' characters.[81] They can save a show from being pulled off the air (the example of *X-Files* and *Jericho*), cause more episodes to be broadcast than originally planned, as in the case of *Freaks and Geeks*, and effectively lobby for the reinstatement of fired actors .[82]

In the September 2007 issue of *The Rolling Stone* there appeared an article in which the author, David Kushner, reports that television producers are increasingly attentive to the voices of online fans.[83] They visit the Web sites devoted to TV shows such as www. TVsquad.com, www. thefutoncritic.com and, most importantly, televisionwithoutpity.com. Television Without Pity is a Web site featuring detailed recaps and often sarcastic reviews of various shows. On its forums users can discuss almost any contemporary American show. 51% of the site visitors are TV professionals, checking the viewers' expectations and reactions to possible changes. For instance, the creator of the *Veronica Mars* teen drama, when working on the show's first season, checked the site in order to make sure that the show's ending would be a surprise for the audience. In March 2007 the site was purchased by Bravo, and executive vice-president of marketing and digital admitted: "We would have built it, if it didn't exist."[84] Online blogging and chat forums on TV shows are a rich and totally free source of information for TV producers; they can learn which trail or character becomes popular with the audience and develop it in subsequent episodes, and they can see which shows should better be canceled and which have a strong fan base. Damon Lindelof, the *Lost* co-creator, argues: "Blogs tell you all the reasons your show sucks," but as the CW executive vice president Michael Roberts admits, "Blogs are immediate, real-time audience testing."[85] Moreover, browsing the Web sites catering to fans, media executives can find talented workers, as it happened in case of some Television without Pity recappers who were hired as writers for *America's Next Top Model* and *Tough Enough* (a wrestling RTV show).[86] Knowing the power of the Internet for fan communication and activism, the more progressive media producers try to

---

[80] Scardaville, "Accidental Activists," 884–85; the author quotes her personal communication with Stephanie Sloane, editor of *Soap Opera Digest*.
[81] Ibid.
[82] Ibid., 887; Jurgensen, "Rewriting the Rules of Fiction;" and Kushner, "TV Enters the Blog World."
[83] Kushner, "TV Enters the Blog World."
[84] Quoted in Kushner, "TV Enters the Blog World."
[85] Both quoted in Kushner, "TV Enters the Blog World."
[86] Ibid.

use it for their own ends. A case in point is the strategy employed by science fiction TV series *Babylon 5* producer J. Michael Straczinski. Communicating through the Internet Straczinsky actively sought the support of fans even before the show was approved for production and used fan support as an argument for the product's market potential.[87]

Conclusion: An Active Audience?

So far describing various fan undertakings and a broad range of creativity, I have painted a picture of an increasingly active audience. Trend watchers aptly noticed the potential of this new type of media consumers and labeled them Generation C, where the C stands for content generated by amateurs who, armed with new, more and more broadly accessible technology, have unleashed their creativity.[88] However, is Generation C a large demographic or only an elite group of media enthusiasts? Opinions vary.

Generation C enthusiasts quote the Pew Internet statistics showing that 44% of US adult Internet users have contributed their thoughts and their files to the net community, 21% declare they have posted photographs on Web sites and 13% maintain their own sites.[89]

"The reigning guru of Web usability,"[90] Jakob Nielsen, draws our attention to different statistics: only 5% of Internet users have weblogs (0.1% users post daily) and 99% of Wikipedia users are lurkers.[91] According to Nielsen, in most online communities user participation follows a 90-9-1 rule, where 90% of users are lurkers, 9% contribute a little and 1% participate a lot and account for most of the action.[92]

Another media expert, Walt Crawford, who as a Net user would certainly qualify as a member of Nielsen's most active 1%, observes that despite much talk about the content generating audience, "Most people, most of the time would rather watch than contribute."[93] After all, even your home YouTube watching time cannot compare to TV watching time.[94]

---

[87] Jenkins, *Fans, Bloggers and Gamers*, 145–46.

[88] trendwatching.com, "Generation C," http://www.trendwatching.com/trends/GENERATION_C.htm (accessed March 24, 2009).

[89] Lenhart, Fallows, and Horrigan, "Reports."

[90] Jakob Nielsen's official Web site.

[91] Nielsen, "Participation Inequality."

[92] Ibid.

[93] Crawford, "Long Live the Audience!"

[94] Ibid.

Henry Jenkins, probably the most renowned fandom scholar, states that the concept of active audience is nowadays generally accepted by everyone in the media industry.[95] Still, he admits that the fans active online are largely white, male, have access to the newest technology and posses the skills to use it. In the US they are also largely middle class and college educated.[96]

Concluding, although the fan communities I have described strongly influence the changing relationships between media production and consumption, they are still an elite audience and an exemplification of how the so-called "digital divide" significantly limits the participatory culture.

Works Cited

Ann Rice's official Web site. Anne's Messages to Fans. IMPORTANT MESSAGE FROM ANNE ON "FAN FICTION." http://www.annerice .com/fa_writing_archive.htm. (accessed March 20, 2009).

Baym, Nancy K. "Talking about Soaps: Communication Practices in a Computer-mediated Fan Culture." In *Theorizing Fandom: Fans, Subculture, and Identity*, edited by Cheryl Harris and Alison Alexander. New York: Hampton Press, 1998.

Brown, Mary Ellen. *Soap Opera and Women's Talk: The Pleasure of Resistance*. Thousand Oaks: Sage, 1994.

Carruthers, Fiona. "Fanfic is Good for Two Things—Greasing Engines and Killing Brain Cells." *Particip@tions* 1, no. 2 (May 2004). http://www.participations.org/volume%201/issue%202/1_02_carruthes _article.htm (accessed March 24, 2009).

Cavicchi, Daniel. *Tramps Like Us: Music and Meaning among Springsteen Fans*. New York: Oxford University Press, 1998.

Chander, Anupam, and Madhavi Sunder. "Everyone's a Superhero: A Cultural Theory of 'Mary Sue' Fan Fiction as Fair Use." *California Law Review* 95 (2007): 597–626.

Crawford, Walt. "Long Live the Audience!" *EContent*, October 2007. http://www.econtentmag.com/Articles/ArticleReader.aspx?ArticleID= 39592&ContextSubtypeID=11 (accessed March 24, 2009).

Friedman, Wayne. "Fan Scripts Touted as New Force in TV." *TelevisionWeek*, December 15, 2003.

Hills, Matthew. *Fan Cultures*. London: Routledge, 2002.

[95] Jenkins, *Fans, Bloggers and Gamers*, I.

[96] Jenkins, *Convergence Culture*, 23.

Jakob Nielsen's official Web site. "About Jakob Nielsen." useit.com. http://www.useit.com/jakob/ (accessed March 24, 2009).

Jenkins, Henry. *Convergence Culture: Where Old and New Media Collide.* New York and London: New York University Press, 2006.

—. *Fans, Bloggers and Gamers: Exploring Participatory Culture.* New York and London: New York University Press, 2006.

—. *Textual Poachers: Television Fans and Participatory Culture.* New York: Routledge, 1992.

Jurgensen, John. "Rewriting the Rules of Fiction." *Wall Street Journal*, September 16, 2006. http://online.wsj.com/public/article/SB115836001321164886-GZsZGW_ngbeAjqwMADJDX2w0frg_20070916.html (accessed March 24, 2009).

Kushner, David. "TV Enters the Blog World." *Rolling Stone*, September 20, 2007. http://www.rollingstone.com/news/story/16697474/tv_enters_the_blog_world (accessed March 24, 2009).

Lenhart, Amanda, Deborah Fallows, and John Horrigan. "Reports: Online Activities and Pursuits." *Pew Internet and American Life Project*, February 29, 2004. http://www.pewInternet.org/PPF/r/113/report_display.asp (accessed March 24, 2009).

Lévy, Pierre. *Collective Intelligence: Mankind's Emerging World in Cyberspace.* Cambridge, MA: Perseus, 1997.

McKinley, Graham E. *Beverly Hills, 90210: Television, Gender, and Identity.* Philadelphia: University of Pennsylvania Press, 1997.

Nielsen, Jakob. "Participation Inequality: Encouraging More Users to Contribute." Jakob Nielsen's *Alertbox*, October 9, 2006. http://www.useit.com/alertbox/participation_inequality.html (accessed March 24, 2009).

O'Connell, Pamela Licalzi. "A World without End for Fans of Jane Austen." *New York Times*, January 13, 2000. http://www.nytimes.com/2000/01/13/technology/a-world-without-end-for-fans-of-jane-austen.html?sec=technology (accessed March 24, 2009).

Pugh, Sheenagh. "The Democratic Genre: Fan Fiction in a Literary Context." *Refractory: a Journal of Entertainment Media* 5 (February 2, 2004). http://blogs.arts.unimelb.edu.au/refractory/2004/02/03/the-democratic-genre-fan-fiction-in-a-literary-context-sheenagh-pugh/ (accessed March 24, 2009).

Sandvoss, Cornel. "One Dimentional Fan: Toward an Aesthetic of Fan Texts." *The American Behavioral Scientist* 48, no.7 (2005): 822–39.

Scardaville, Melissa C. "Accidental Activists: Fan Activism in the Soap Opera Community." *American Behavioral Scientist* 48, no. 7 (2005): 881–901.

Scodari, Christine, and Jenna L. Felder. "Creating a Pocket Universe: 'Shippers,' Fan Fiction, and *The X-Files* Online." *Communication Studies* 51, no. 3 (2000): 238–57.

Shaw, Suzy. "The Internet as an Entertainment System." *Bulletin of the American Society for Information Science* 21, no.1 (1994): 9–11.

Thomas, Lyn. *Fans, Feminisms and "Quality."* London: Routledge, 2002.

Thornton, Sarah. "The Social Logic of Subcultural Capital." In *The Subcultures Reader*, edited by Ken Gelder, 186–87. New York: Routledge, 2005.

Walker, Andrew. "Harry Potter Comes of Age, and Other Stories." *BBC News Online*, August 29, 2003. http://news.bbc.co.uk/1/hi/magazine/3188813.stm (accessed March 24, 2009).

Williams, Rebecca. "'It's About Power': Spoilers and Fan Hierarchy in On-Line *Buffy* Fandom." *Slayage: The On-Line International Journal of Buffy Studies*, no. 3.3–4 (April 11–12, 2004). http://slayageonline.com/essays/slayage11_12/Williams.htm (accessed March 24, 2009).

Young, Cathy. "The Fan Fiction Phenomena: What Faust, Hamlet, and Xena the Warrior Princess Have in Common." *Reason* 38, no. 9 (2007): 14–15.

HUNTING FOR RACISM—ONLINE CITIZEN CRITICS' REVIEWS OF KEVIN MACDONALD'S *THE LAST KING OF SCOTLAND*

ANNA KRAWCZYK-ŁASKARZEWSKA

This paper will be devoted to some of the practices and strategies adopted by the so-called online citizen critics who post their non-expert opinions on Web sites devoted to film and film-making. *The Last King of Scotland*, a film in which the themes of power, race and history have been interwoven to create a fascinating and controversial tale, provides a fairly representative example of how online amateur "journalism" is trying to catch up with professional reviewing and undermine its cultural hegemony, while confirming the social stereotypes it claims to demystify. The deliberately ambiguous phrase: "hunting for racism" in the title of the present article should be a reminder that the "racist beast" is still alive and well in the modern-day world, fed by human ignorance and malevolence, but in the process of looking for it the searchers often run the risk of picking out and concentrating on the wrong target.

Facts and Fictions

The Last King of Scotland is a 2006 film adaptation of Giles Foden's novel of the same title, scripted by Peter Morgan and Jeremy Brock, and directed by Kevin Macdonald. The protagonist of the story is Nicholas Garrigan, a young Scottish doctor who has just graduated from medical school and, in order to escape the stifling predictability of his homeland and the professional dominance of his father, chooses to go to Uganda, where he hopes to "make a difference" by helping the people in need, and to have some fun. It is 1971 and Garrigan's arrival in Uganda coincides with the military coup which brings General Idi Amin to power. The rash and arrogant youth does not even bother to learn anything about the history and the political situation of the country. The new leader's speech, delivered during a political rally, greatly impresses him, although he is warned by his co-worker Sarah Merrit, a white female doctor, that the cheering crowd "sang just like that for Obote, until they realized that he turned their economy into his personal bank account." As a result of an unexpected road encounter with the general, Garrigan accepts an offer to be Amin's personal physician and soon becomes his aide and closest confidant. Garrigan's hedonist naiveté, his refusal to believe the warning signs from British diplomats (whom he accuses of an imperialist agenda) and intoxication with power prevent him from seeing through Amin's clownish facade. He is foolish enough to cuckold the dictator by seducing and impregnating Kay, one of his neglected wives. When he finally realizes Uganda is ruled by a murderous, paranoid tyrant, his own life is in danger. He manages to save it only thanks to the sacrifice made by one of his colleagues, Dr Junju, a black man who has no illusions concerning Amin.

One of the major problems as far as the reception of *The Last King of Scotland* is concerned, results from the peculiar advertising strategy preceding its release. The film was marketed mainly as a biopic of the recently deceased dictator, yet it introduced fictitious characters and modified at least some of the historical events accompanying the rule of Idi Amin. In fact, numerous viewers/critics complained about the difficulty in distinguishing what was accurate historically and what was not. It also seems that the expectations of the audiences concerning the presentation of Amin were based on diverse, mutually incompatible agendas. Some of the comments, reviews and essays written in connection with the film, tended to blame it for perpetuating stereotypes of the figure of the corrupt African dictator, others claimed the truth about Amin's reign was much darker; there were even opinions suggesting that *The Last King of Scotland* provided a shameless glorification of a politician who, in reality, was a ruthless and corrupted murderer, and that it conveniently omitted the issue of Great Britain's complicity in installing Amin into power.

Kevin Macdonald's background as a documentary filmmaker (the award-winning *One Day in September* and *Touching the Void*) to a certain extent helps understand some of the strategies employed in *The Last King of Scotland*, both in terms of the story and its visualization, but certainly does not diminish the controversy stemming from the fact that the story of Ugandan misery revolves around a fictional white character who tampers with real people and events.[1] Perhaps at the core of the debate is the issue of whether movies should be educational or entertaining, or both. Rather

[1] Nicholas Garrigan is, in fact, a composite character based on three authentic people who worked for Idi Amin (a fact mentioned in numerous reviews of the film and interviews with its director).

surprisingly, at least from the point of view of commercial mainstream filmmaking, the reception of *The Last King of Scotland* was imbued with ethical considerations. Shot on location in Uganda, at times using authentic props and sites visited by Amin and his entourage, the film was expected by many to provide an accurate portrayal of the dictator's rule, and when it turned out that Amin's story was treated as secondary, indignation and search for culprits ensued. One optimistic conclusion to be drawn from the situation is that, apparently, films still matter, even if the majority of them constitute mindless, populist entertainment. On the other hand, the nagging questions remain and recur: how much truth do we have the right to expect as contemporary cinema goers, what level of complexity should be achieved while dealing with a nation's past trauma, in a cultural artefact whose primary goal is to entertain the viewers?

Hollywood Does Africa

In his seminal study *On Visual Media Racism* (1978), Eugene Franklin Wong catalogued the stereotypical representations of Asian characters in American motion pictures. With minor modifications, the list could very well be applied to the way in which Hollywood tends to portray people and issues connected with the African continent. Among the frighteningly up-to-date items of Wong's list one could mention the primacy of the (sympathetic) white point of view, the fact that black characters appear in secondary plots and are mostly one-dimensional, static, either monsters or victims, whereas white characters are given starring roles and a chance to develop; promiscuous black women who are always sexually available, as opposed to white women who have a degree of control over their fates and can sometimes be shown as virtuous, etc. "African" films such as *The Constant Gardener, The Interpreter*, *To Kill a Mockingbird*, *Blood Diamond*, *Amistad*, *A Time to Kill* or *Cry Freedom*, share a rather disturbing propensity, namely they concentrate on the white experience and the beneficial role of the white protagonists in the lives of helpless, strife-torn black communities. Joe Queenan reduces this persistent tendency to the mocking "White Folks to the Rescue" title and, while reviewing *Blood Diamond*, he comments ironically on the necessity of introducing the deeply caring, savior-like white characters who always manage to conquer evil, even if it is inflicted by other, not-so-noble white characters: "If there is anything black people the world over have learned from Hollywood—and there isn't a whole lot—it's that no matter how bleak the situation seems, they can always rely on some resourceful, charismatic and, in some instances, shapely white person to bail them out."[2]

In accordance with Spike Lee's ironic remark, *The Last King of Scotland* has been categorized by many critics and reviewers, both professional and nonprofessional, as a film which "goes into the growing genre of 'movies about black people that end up being about white people.'" Gabriella Beckles went as far as to claim that "*The Last King of Scotland* epitomized all that is wrong with Hollywood's vision of Africa."[3] However, the most damning accusations against *The Last King of Scotland* were amassed by Vanessa Walters, who—even before Forest Whitaker, the actor playing Amin, won an Oscar for his role—decided that if he did, it would be because, just like in the case of Halle Berry and Denzel Washington, "once again Hollywood's racist beast has been duly sated."[4] Furthermore, Walters accused the creators of the film of presenting the African continent in a blatantly clichéd way in order to sell it to the white audiences:

> Amin is believed to have killed at least 300,000 people—but does that give carte blanche to the film-makers to play to some of the worst stereotypes of corrupt, murderous, incompetent and ridiculous black leaders? . . . Most reviewers have failed to clock the fact that the "white man trying to save ethnic man" from himself is a well-worn caricature. Nor do they question the other easy stereotypes that crop up in the film. Apart from the scary men running around Uganda with AK47s, there is a plethora of scantily clad go-go dancers and other exotic, sexually available women to be bedded by Garrigan—including one of Amin's wives. . . . There are flies on sick, starving faces, and the hospitals are filled with filthy swarms of humanity. There is the selfish, corrupt elite in their manicured compounds and the peasants portrayed as helpless children, making an unintelligible warble in, y'know, "their language." Africa is presented as a place of violence and superstition, ruled by fear.[5]

Taking into consideration the history of "Hollywoodized" Africa on celluloid, the blogger's arguments seem to make perfect sense, yet in the context of *The Last King of Scotland* they appear strikingly superficial, if not altogether misguided, not only because the author of the above passage did not bother to consider the validity of at least some of the clichés she mentions. Macdonald's film cleverly touches upon numerous Africa-

[2] Queenan, "Whiter Shade of Guile."
[3] Beckles, "Hollywood's Vision of Africa."
[4] Walters, "Stereotypes That Will Sell."
[5] Ibid.

related stereotypes and reverses practically all of them. The Scottish doctor is not shown in a sympathetic way; although he is the protagonist, his arrogance, selfishness and mind-boggling stupidity prevent the audiences from identifying with his plight. "That Nicholas—however inadvertently, however much he seems a victim—is also capable of great horrors, specifically in that he lets others act it for him and gets away with it (emotionally, morally, politically, if not physically), makes him, at some deep level, even more troubling than Amin."[6] Idi Amin, in turn, strikes the viewers as a multi-faceted character: buffoonish, but never reduced to a caricature; unpredictable, but very human even when subject to bouts of fury; ruthless and paranoid, but also warm-hearted, capable of compassion. As far as the sexual availability of women in the film is concerned, Garrigan has paid sex with one black girl and "beds" Amin's estranged wife: a feat which suggests either Garrigan's utter lack of brain or the film scriptwriters' tendency to stretch the credibility of the story, but can hardly be classified as impressive or racist.

Jan Stuart's objections are of an even larger caliber. He insists *The Last King of Scotland* is a "shamelessly manipulative melodrama" in the colonialist vein and "a return to the pandering celluloid depictions of African turmoil that insist on putting a white face on black suffering": "Hundreds of thousands of Ugandans are being slaughtered, but we are not allowed to feel the full force of Amin's brutality 'til it hits our man from Scotland. I haven't felt so soiled at the movies since Roberto Benigni co-opted the tears of 6 million in *Life Is Beautiful*."[7] On closer examination, however, it becomes rather obvious that since *The Last King of Scotland* consistently maintains Garrigan's point of view, it is bound to reflect his narrow-mindedness and ignorance concerning the suffering of the Ugandan nation. All in all, one cannot help the impression that Macdonald's film has been unfairly lumped with other Hollywood productions portraying African countries and people from the white man's perspective. While assessing it, critics/reviewers often resorted to stereotypes themselves, because they assumed a western-made movie about Africa must be racist.

At the expense of historical accuracy, the viewers are offered a cautionary tale of the seductive lure of power, in which the only truly noble character is a black Ugandan doctor who sacrifices his life in order for Garrigan to be able to flee the country and tell the world about Amin's atrocities. The ultimate irony is that the African "adventure" does not

---

[6] See Fuchs's review of *The Last King of Scotland*.

[7] Stuart, "*Last King*."

teach Garrigan anything. He begins his journey as a clueless man. At the end of the story he remains clueless, and very scared: he only wants to save his skin. *The Last King of Scotland*, both the novel and its film version, might then be conceived as an ironic take on the genre exemplified by *Heart of Darkness* and *Moby Dick*, and defined by Paul Zweig as "new adventure myth":

> This new adventure myth portrayed nineteenth-century Western culture, with its emphasis on "domestic values"—the primacy of work, family, and obedience to law, conscience, and morality—as a prison from which adventure offered escape. The hero of this new adventure myth was a rebel against the established order who escaped the realm of domestic values into a realm beyond conventional morality, "beyond good and evil."[8]

For Garrigan there is no follow-up, no moral transformation or redemption. However, the universal mechanism of accepting the dictatorial modus operandi by those who are subjected to the dictator's rule, and who want to reap benefits of the diabolical pact with the evil power, has been revealed here with an uncomfortable precision.

Citizen Critics React

Taking into account user comments and customer reviews offered by the so-called "citizen critics"[9] on Web sites such as IMDb, Rotten Tomatoes or Amazon.co.uk, it will be no exaggeration to state that the film provoked extreme reactions not only after it was shown in cinemas, but even before it was released. The first signs of discontent were indicative of a general mood concerning Hollywood films about African issues: scepticism and mistrust prevailed. *The Last King of Scotland* was almost immediately classified as a Hollywood product by most posters on the IMDb message board devoted to the film, although it would have been more prudent to treat it as an independent production, since it was financed by British and German money, and distributed by Fox Searchlight, a branch of the big Fox corporation which is usually interested in low budget films, rather than movies made for blockbuster appeal.

When the film hit the cinemas, nearly all imaginable kinds of reservations and objections were voiced. According to some of the threads

---

[8] Whaley, "Adaptation Studies," 38.

[9] After Rosa Eberly, I use the term in a general sense, to refer to nonprofessional expertise.

originated by citizen critics on IMDb, *The Last King of Scotland* was, in turn, anti-English, anti-Scottish, anti-Ugandan, racist, misogynistic, too politically correct, PC-insensitive, a piece of propaganda pandering to white liberals, a piece of propaganda meant to satisfy white and/or Asian reactionaries, a piece of propaganda aimed at extolling the virtues of a black barbarian, and so on and so forth. Instead of analyzing the subtleties and deficiencies of the script, the "civilian" critics opted for sweeping statements, e.g. "*The Last King of Scotland* is the most racist, propaganda film I've seen in the last ten years," or "This is Hollywood's attempt at creating a Black Hitler, since none exists." Worse still, the opponents of the film suggested Whitaker's portrayal of Idi Amin was one-dimensional; in a classic post-Oscar backlash, reminiscent of some of the professional rants quoted above, they argued it was only because the actor naively agreed to take the "black face" role and present Amin as a ridiculous, blood-thirsty despot that he could be awarded an Oscar by the members of the Academy. At the other pole were opinions conveniently omitting the questionable choices made by the writers and crew of the film and hailing it as a groundbreaking masterpiece: *The Last King of Scotland* "closes the door on sentimental African adventure movies once and for all. Without preaching and without stereotyping for once it has been done well."[10] To cut a long story short, there were very few shades of grey in this black-and-white exchange.

Occasionally, however, it was possible to trace a different, slightly subversive tone, reminiscent of the tradition of grotesque hyperboles used by satirists such as Jonathan Swift or Ambrose Bierce. For instance, in the IMDb thread commenting upon the necessity of casting white actors in *The Last King of Scotland* so as to attract white audiences, one of the posters reacted in the following manner:

> Personally, I was disappointed at how many black people were cast in this movie. I was hoping to see more white people in black body paint. Fortunately, there were just enough white people in the film to redeem it.[11]

Particularly interesting were the attempts to either "disarm" the accusations of the alleged or real racism or, conversely, emphasize the racist tricks. Anger, sarcasm, bitterness, irony, lectures in history, insults,

---

[10] "*The Last King of Scotland*," comment posted February 1, 2007, http://cobb.typepad.com/cobb/2007/02/page/2/.

[11] A response by poster nicknamed trumpydumpy2001, in the thread titled "Face it, if there were no whites in the movie," posted June 10, 2007 on the Internet Movie Database (IMDb), http://uk.imdb.com/.

calls for boycott, detailed sociopolitical analyses, and, predictably, insults and threats: all these ingredients could be found on a message board whose content was, by definition, provided by nonprofessionals for the purpose of entertaining rather than taking a firm political stance. The following fragment of the "Hello? Racism" thread constitutes a representative sample of how online debates are carried out in connection with sensitive issues, and how difficult it is to establish even a basic understanding among people of different races, educational backgrounds, and motivations:

> I can't believe how stupid people can be to like this movie. I consider myself to be something of a movie buff with my favorites ranging from the "Rocketeer" to "Stomp the Yard" but I have never seen a movie with such overtly racist undertones. I should know because I am in AP english and the whole class is devoted to our teachers telling us what to infer about the plot of really long books. It is really cool, it's pretty much just us white kids, a couple asians and I think there is one jew, also.
>
> I think we can all agree that there are many racist, Nazi-like individuals who claim that all African-Americans look the same. I know from experience that this is not true because we have like 4 that go to my school. they are bussed in from the poor areas and only 2 of them look the same but that is because they are brothers . . . like, actual brothers, not just black people. What I mean is that they have the same mother . . . and possibly the same father.[12]

The above text performs several functions: it mocks not only racism, but also the desperate search for racism; it parodies the one-sidedness of online rhetoric and the very way in which citizen critics pass their verdict on films. It also pokes fun at the educational systems of the contemporary Western Europe and the United States, which offer ready-to-use scripts for identifying racism in cultural artefacts, but do not seem to provide enough historical background to make the act of recognition truly significant. Ironically enough, the grotesquely titled thread started by bengati83 continues to be taken seriously and attract indignant or patronizing comments. Quite apart from the obvious conclusion that, perhaps due to linguistic and/or cultural differences, some of the posters did not realize the original post was meant to be ironic, one needs to question the extent to which the reservations voiced by the posters are legitimate. In other words, it does not seem entirely fair to dismiss the aggressive responses as mere evidence of hypersensitivity or excesses of Political Correctness.

[12] The "Hello? Racism" thread started by bendigati83, posted June 9, 2007, the Internet Movie Database (IMDb), http://uk.imdb.com/.

Perhaps a more rational explanation could be that the viewing public internalized the idea that Hollywood's representations of Africa are inherently racist and approached *The Last King of Scotland* as yet another instance of race-related bias. Shallow and convenient? Certainly. It should be remembered, however, that many professional critics and reviewers were not interested in an in-depth analysis of the film, either, contenting themselves with labelling it as a typically racist Hollywood product.

Professionals Strike Back

In her paper on amateur book reviewing, Lisa Ede expresses the belief that "online citizen reviews represent the first major challenge to the traditional system of print reviews since this system appeared in the 18th century." Ede is convinced ordinary citizens' creative output makes it possible for them not only to "share their views with anyone who has a computer but also to challenge the norms of high culture and the authority of conventional expertise."[13] Her remark might serve as a useful point of reference in examining the consequences of the overwhelming proliferation of nonprofessional film journalism and/or criticism. At the risk of simplification, it could be argued that, similarly to the issue of race representations, the status and importance of online citizen critics provokes extreme reactions and unfair generalizations, particularly on the part of those film critics and reviewers who started their careers in the traditional media. The abundance of amateur writing about film will, then, be treated by some of those participating in the debate as a genuine challenge to the hitherto prevalent cultural hierarchies, an unprecedented opportunity to deal with the limitations of the new digital democracy, or, conversely, it will merely confirm the ultimate and irreversible dumbing-down of the contemporary audiences.

The somewhat uncertain status of the film critic in the age of the Internet has been a subject of even more heated debates, with new arguments added by well known and respected film bloggers practically every few weeks. Todd McCarthy's pessimistic pronouncement that professional critics will soon be jobless, because "[t]he age of thoughtful, considered, in-depth artistic evaluation is over, the argument goes, and the time of quick-draw, highly personalized, instantly disseminated snap judgments has arrived,"[14] created much controversy among his equally popular colleagues, e.g. Joe Morgenstern or David

[13] Ede, "Online Citizen Book Reviews."
[14] McCarthy, "Critics on Criticism."

Poland, for whom the democratization of the process of reviewing represents a fascinating opportunity, rather than a reason to be frustrated or apprehensive. On the other hand, although McCarthy's diagnosis, according to which online users are exposed to "a free-for-all, an opinion consumer's wet dream, a bazaar of commentary that constitutes the critical-world version of a global free-trade agreement,"[15] sounds menacing, it at least encourages specific questions concerning the desired role of film critics nowadays: what level of authority should they aim for, should their reviews stimulate discussions, e.g. about the state of the film industry or social inequalities, or should they merely provide the reader with reliable consumer-related information?

In response to McCarthy's gloomy ruminations, Poland argues that "the Number One reason for the diminishment of the critic has nothing to do with the web or bad decisions by editors or box office chatter . . . it's marketing dollars. Period. End of discussion." More importantly, he comments on the seemingly unproblematic advantage of online writing over the print media, namely the faster content delivery: "Speed is easy. Any monkey can do something fast. Internet audiences are built on engagement."[16] Ultimately, and not surprisingly, the issue boils down to the quality of the content offered to the readers. Rather than worrying about the "diluted" power of individual critics (which is, perhaps, McCarthy's subtle code for critics of renown), it will be more productive to admit that on the whole the alleged power of the established film critics has rarely been a crucial factor in the decisions made by film viewers, and that the stronger presence of "passionate and eloquent movie lovers"[17] who happen to be outsiders from the point of view of film industry, finally offers the potential for creating a multilayered critical space in which the Web-savvy audiences are not forced to oscillate between the "old school" delusional elitism and ostentatiously corporate-friendly consumer guiding.

Works Cited

Beckles, Gabriella. "Hollywood's Vision of Africa: The Roles of Black Folk in Film." *Suite 101*, May 12, 2007. http://racism.suite101.com /article.cfm/hollywoods_vision_of_africa.

Eberly, Rosa. *Citizen Critics*. Urbana: University of Illinois Press, 2000.

[15] Ibid.

[16] Poland, "Critics on Criticism on Criticism."

[17] Morgenstern, "Rumors of Critic's Demise."

Ede, Lisa. "Online Citizen Book Reviews and the Circulation of Cultural Power." Paper, ISHR Conference, July 16, 2005, Los Angeles, CA. http://thewritingway.blogspot.com/2005_07_01_archive.html.

Fuchs, Cynthia. "Playing the White Man." Review of *The Last King of Scotland*, directed by Kevin Macdonald. *Popmatters*, September 27, 2006. http://www.popmatters.com/pm/review/the-last-king-of-scotland-2006.

Julian Walker's official Web site. "*The Last King of Scotland*: Post-colonial Complexities." Review of *The Last King of Scotland*, directed by Kevin Macdonald. May 21, 2007. http://julianwalkeryoga.gaia.com/blog/2007/5/the_last_king_of_scotland_post-colonial_complexities.

McCarthy, Todd. "Critics on Criticism: Film; Blogosphere Changing the Role of the Critic." *Variety*, October 12, 2007. http://www.variety.com/article/VR1117973973.html?categoryid=2520&cs=1.

Morgenstern, Joe. "Rumors of Critic's Demise Are Greatly Exaggerated: The Role of Film Reviewers in an Internet Age." *Wall Street Journal*, April 29, 2006. http://online.wsj.com/public/article/SB114625940532139146-C7jmprLOOjx1SCIG3LrAxhQgGoc_20070430.html?mod=tff_main_tff_top.

Poland, David. "Critics on Criticism on Criticism: Film; Blogosphere Being Used as Scapegoat in Changing Role of Traditional Critics." *MCN*, October 13, 2007. http://www.moviecitynews.com/voices/2007/071013_poland_mccarthy.html.

Queenan, Joe. "A Whiter Shade of Guile." Review of *Blood Diamond*, directed by Edward Zwick. *Guardian*, January 5, 2007. http://www.guardian.co.uk/film /2007/jan/05/leonardodicaprio.

Stuart, Jan. "*Last King* is Not What the Doctor Ordered." Review of *The Last King of Scotland*, directed by Kevin Macdonald. *Newsday*, September 29, 2006. http://www.newsday.com/entertainment/movies/ny-etopen4909013sep29,0,3623109.story?coll=ny-movies-utility.

Walters, Vanessa. "Stereotypes That Will Sell: *The Last King of Scotland* Is Not the African Idi Amin's Story; It Is a White Story for White Audiences." Review of *The Last King of Scotland*, directed by Kevin Macdonald. *Guardian*, January 18, 2007. http://www.guardian.co.uk/commentisfree/2007/jan/18/comment.uganda.

Whaley, Donald M. "Adaptation Studies and the History of Ideas: The Case of *Apocalypse Now*." In *The Literature/Film Reader: Issues of Adaptation*, edited by James M. Welsh and Peter Lev, 35–50. Lanham, MD: Scarecrow Press, 2007.

Wong, Eugene Franklin. *On Visual Media Racism: Asians in the American Motion Pictures*. New York: Arno Press, 1978.

NEGOTIATING THE VIRTUAL: *THE MATRIX*, THE INTERNET AND A NEW TECHNO-LOGIC

MAREK M. WOJTASZEK

> Everything is a machine . . . There is no such thing as either man or nature now, only a process that produces the one within the other and couples the machine together . . . Everything is production.[1]
> —Gilles Deleuze and Félix Guattari

Abstract: The Internet has by now become a commonplace, a major structuring component of our reality, which both epitomizes the velocity of technological changes and symbolizes the stage of civilizational development. Even though it was originally conceived and constructed to facilitate people's works and lives, its service potential has largely been colonized by capitalism and harnessed to its policies and strategies whose exploitative character often remains invisible. This capitalistic, and indeed cultural, figuration and use of the virtual space continues to perpetuate the dyad of the human versus the machine, whereby the former is valued to the detriment of the latter, always posited as secondary and inferior. The poststructuralist endeavor to dissolve the original, exposing it as a mere illusion whose authenticity has been constituted through iteration amounts to a simple reversal of terms, whereas dualistic structure is sustained. Consequently, we encounter cultural representations of machines in the service of humans, or machines seizing control over humanity. Through an analysis of *The Matrix* the paper exposes this binarized framework as incapable of thinking the critical potential of thc virtual, of which the Internet is a real world example. Drawing upon the works of Gilles Deleuze and Félix Guattari, the paper proposes a novel rendering of the Internet along the lines of their fundamental concepts of the rhizome,

[1] Deleuze and Guattari, *Anti-Oedipus*, 2, 4.

immanence and the virtual. This manner, overcoming the distinction between reality and virtuality, signals the demise of the rationalistic legacy and the humanist bias inscribed in this. It enables a radical rethinking of the world and the role of the Internet in creating the social imaginary. The emerging new techno-logic, it is argued, is better suited to accounting for global and social changes as well as for the significance and potential that the Internet offers to post-humanity and its new reality of the virtual.

Introduction: Beyond Negativity

The second half of the twentieth century undeniably witnessed an exceptional surge and development of divergent technologies. Interestingly, their intensified growth and popularization coincided with the resuscitation of politically engaged thought within a larger school of equally diversified thinkers who came to call themselves poststructuralists. Indeed, even though not unprecedented in Western history, this period saw a unique alliance between *techne* and *episteme*. The Greek beginnings of Western thought, embodied in the triad of Socrates, Aristotle and Plato, laid the groundwork for the conceptualization of technology for many centuries to come. All three viewed *techne* (craft or art) at worst as threatening to the rationally established peace, order and governance (Socrates, *Ion*), proof of human imperfection and poverty embedded in the human desire to imitate nature (Aristotle) or, at best, as secondary to the knowledge of forms (Plato, *The Republic*). All in all, for ancient Greeks *techne*, construed as a broad term encompassing all the mechanical arts, was defined in strictly negativistic terms, which no doubt affected its social and cultural positioning. Ever since, due to this view, it has only been considered fit for the lower classes, whilst the upper class representatives were busy practicing the liberal arts, designed for "free" men.[2] *Techne*'s primary task of applying a method, accomplishing, effectuating—i.e. of production itself—was relegated to the auxiliary one, which must always be anteceded by the original work of the faculty of the understanding. Put differently, reversing Nietzsche against himself, there is always being behind doing. Production is doomed to mere subservience and inferiority. Thus, the technological other played a constitutive role in the sustaining and functioning of Western dualism. The obvious monstrosity of technology has forever accompanied human development and shaped their imagination. This process of co-existence proved to abound in varied representations of technologies, ranging from its

[2] Dorter, "The Ion," 65–78.

deification to utter brutalization, always exciting polarized feelings of either fascination or revulsion. "Our machines are disturbingly lively and we ourselves are frighteningly inert," as Donna Haraway describes the contemporary state of art in the relation between the human and the technological.[3]

It is with the assault on the Greek foundations of Western thought (metaphysics and epistemology) leveled by the aforementioned poststructuralist thinkers (Lacan, Irigaray, Lyotard, Butler, Derrida, Bourdieu, Deleuze) that one can observe a radical shift in the rendering of technology and its potential. The nearly two-millennium-old ontological dualism of man versus machine could now be subject to an incisive scrutiny and be thoroughly reconsidered. Technology found itself freed from the constrictive corset of rationalistic explanations and was given new space for self-expression and experimentation. Its radical potential, not exclusively in the practical field of application, but also on the level of theory, was very rapidly discerned and put to use. Here, the outstanding cases of thinkers such as Paul Virilio, Michel Foucault, Bruno Latour, Donna Haraway and many others testify to the immense scholarly interest and proliferation of writings on the technological. This academic investment, as a matter of fact, runs along the same old lines of fascination with the powers of technology; however, this time the intention is to underscore the permeability of the extrapolated oppositions of man and machine, and, consequently, to blur the ontological boundaries separating them, to the point of their self-explosion. The motivation behind this rather long introduction tracing the theoretical base of technology is not solely to emphasize its original negative status but, most importantly, to point in the direction of novel, more affirmative renditions.

Of course, one major gate for experimentation with the technological was opened by the development of computer-based, digitalized technology and what came to be known as "cyberspace."[4] What used to be the ancient world of magic, myth and legend as well as the modern world of fantasy, fiction, movies and cartoons has now become artificially generated (unreal) realities, of which cyberspace in the form of the Internet is a real world example. Gibson in his *Neuromancer* reads "cyberspace" as,

> [a] consensual hallucination experienced daily by billions of legitimate operators . . . A graphic representation of data abstracted from the banks of every computer in the human system. Unthinkable complexity. Lines of

[3] Haraway, *Simians, Cyborgs and Women*, 152.
[4] Gibson, *Neuromancer*.

> light ranged in the nonspace of the mind, clusters and constellations of data.[5]

This new dimension created by humans and technology violates the most rudimentary notions structuring our reality, i.e. the Kantian transcendental principle of time (a condition of perception imposed by our own mind) and the Euclidean conception of space (virtual space as opposed to physical space). Moving beyond the strictly material and empirical, it offers a revolutionary method of understanding the basic categories of humanity, perception, and experience. This opening to the virtual accomplished via technology has made humans face again the perennial question of images, which takes us back to our starting point, i.e. ancient Greece. Therefore, prior to excavating the radical, and affirmative, potential of the virtual as it is exercised by the Internet, I will, through a critical reading of *The Matrix,* adjudicate the role images played in Western thinking and reassess their impact on human existence. In order to explicate their involvement in the maintenance of the pervasive ontological dualism, and thus to understand the peculiar proximity and similarity that they bear to technology, a brief return is necessary, *ad fontes,* to the founding father of the Western theory of images, that is Plato.

"Free your mind": Dissimilarity and Generativity

> What is the Matrix? Neo inquires.
> It's that feeling you have had all your life. That feeling that something was wrong with the world. You don't know what it is but it is there, like a splinter in your mind, driving you mad . . . The Matrix is everywhere, it's all around us, even now in this room you can see it when you look out of the window, when you turn on your TV. You can feel it when you go to work, when you go to church, when you pay your taxes. *It is the world that has been pulled over your eyes to blind you from the truth*, Morpheus replies.[6]

What kind of a world then precludes our experience of the Matrix and what, more precisely, the Matrix *really* is, seem to surge to the fore as the two most essential questions. They will, therefore, guide our deliberation about the significance of images and their structuring force within our reality. Reading Plato against himself, Gilles Deleuze detects another

[5] Ibid., 51.
[6] Wachowski and Wachowski, *Matrix* (my emphasis).

instance in-between what Plato posited as the rigid dualism of the Real and the copy.[7] This element—"an image without resemblance"[8]—pulverizing the fundamental and unconditional principle of similarity, which is assumed between the form and appearance, is retrieved and awarded a primary status. Pure image—a simulacrum—is affirmed as a radical force which immediately signals the beyond of ontological dichotomy of essence and representation. Instead then of viewing the actual (real) world as distinct from (virtual) images, or metaphors, which we utilize to re-present it, thus perpetuating the intrinsically Western metaphysical bias, Deleuze introduces the power of imaging—the virtual—as primordial, as effective of the actual, and in so doing he postulates a strictly monistic vision of the real. Deleuze maintains:

> The simulacrum is not a degraded copy. It harbors a positive power which denies the original and the copy, the model and the reproduction. At least two divergent series are internalized in the simulacrum—neither can be assigned as the original, neither as a copy.[9]

Not only is the uniqueness of the simulacrum posited in terms of its singularity and dissimilarity from both the original and the copy, but also its intrinsic in-formality. As Leonard Lawlor writes, "In-formality means that since the simulacrum begins with the formless (chaos), its repetition is always unformable (different), subject to events."[10]

This is not in the least an anti-realist theory. Deleuze does not claim that reality is a mere image or is constructed by the mind. One cannot encounter here the famous Baudrillardian lament for the real; rather, in a uniquely Dionysian act of destruction of the world as we know it, which Deleuze describes as representation, one welcomes the ultimate "twilight of the idols." Hence, in all its difference and complexity, reality cannot be limited to extended images which humans have *form-ed* of it. Neither can the mind be viewed as the author or origin of all images. On the contrary, the real is an incessant, infinite and inhuman plane of imaging that knows no resemblance or crude oppositions between the Same and the Similar, themselves being only one of numerous possible outcomes of the productive—indeed technological—machinery of the simulacrum. Nowhere is this better illustrated than in the scene of Neo's literal metamorphosis. Seated by the mirror, he is watching his self-image

---

[7] Deleuze, "Simulacrum and Ancient Philosophy."
[8] Ibid., 295.
[9] Ibid., 299.
[10] Lawlor, "Beginnings of Thought," 68.

reflected in the mirror-machine; gradually he realizes his power to create images, his simulacral creativity, and in an act of reunion with the mirror he melts down, overwhelmed by the fluidity of imagistic form, he is becoming-image himself. Further, Deleuze clarifies that "[b]y rising to the surface, simulacrum makes the Same and the Similar, the model and the copy, fall under the power of the false (phantasm)."[11] Now, it is clear that we are leaving the old metaphysical traditions of substance behind, both those of idealism (the idea produces the world), materialism (the world produces the idea) and their poststructuralist negativistic reconfigurations (there are only images, the real ceases to exist) as well as all Apollonian transcendence (subjectivism, otherworldliness). Conversely, the loss, the infamous Baudrillardian echo in the film—"the desert" of the real heralds a new beginning, a resurrection to the *real*(m) of the unknown, yet unexplored beyond the Platonic dyadic rendering; a threshold of immanence. As Deleuze remarks, "The desert, experimentation on oneself, is our only identity, our single chance for all the combinations which inhabit us."[12] The real is far richer than that negativistic legacy implies.

In its allegedly all too unrealistic, oneiric or phantasmatic vision, *The Matrix,* bringing in the technological *other* and emphasizing its generative potential, excels in overcoming the dualism of human versus machine which ensues from the representational mode of thought. Therefore, in no way can the eponymous "matrix" be poorly understood in the psychoanalysis-driven manner as that advocated by Žižek.[13] Far from being a metaphor of what—he suggests—Lacan called "the big Other," the virtual symbolic order, the network that structures reality for us, the simulacral "matrix" emerges as purified of all negativity and is welcomed as pure womb-like virtuosic generativity, whose portrayal in the form of all-technological generation is well articulated in the film.

> Throughout human history we have been dependant on machines to survive. Fate, it seems, is not without a sense of irony. The human body generates more bioelectricity than a 120 volt battery. Combined with a form of fusion, the machines have found all the energy they would need. There are fields, endless fields where human beings are no longer born. We are grown by machines, Morpheus explains.[14]

---

[11] Deleuze, "Simulacrum and Ancient Philosophy," 300.
[12] Deleuze and Parnet, *Dialogues II*, 9.
[13] Žižek, "Matrix."
[14] Wachowski and Wachowski, *Matrix*.

It is clear that the body itself is a machine, of which *The Matrix* never ceases to remind us through a multiplicity of sockets, orifices and plugs with which the body is equipped, as well as serpentines of cable by means of which it connects with other bodies. It is nothing but millions of machinic processes and connections which are productive and repetitive. Repetition proceeds by the eternal recurrence of the same, the only same that can ever return—i.e., the different.[15] "It is not even enough to invoke the model of the Other, for no model can resist the vertigo of the simulacrum," Deleuze stresses.[16] Put otherwise, any attempt to *form*-ulate identity from the abyss of the Other will necessarily be condemned to a failure precisely because this dualistic framework is all it is—a limitation, a product, an effect that only achieves its original status through representation, in an act of paranoid inversion of order between cause and effect.

Implicit in this is again an appeal to reconsider our relation to the Other, the Outside, no longer as a simplistic negation, absence of order and regimented binary difference; rather through affirming it as multiple productive differences that have not yet been encountered, thought, or sensed. Brian Massumi flatly states, "The challenge is to assume this new world of simulation and take it one step further, to the point of no return, to raise it to a positive simulation of the highest degree by marshalling all our powers of the false toward shattering the grid of representation once and for all."[17] Swallow the (red) pill, flee to the dream world, let yourself be lulled and be led by Morpheus to Wonderland, where the power of the false will cause your metamorphosis, as it is able to transform itself according to the forces that it encounters. Morpheus asks, "Have you ever had a dream, Neo, that you were so sure was real? What if you were unable to wake from that dream? How would you know the difference between the dream world and the real world?"[18]

"Follow the white rabbit": Deterritorialization

> "What truth?" Neo asks.
> "That you are a slave, Neo. That you, like everyone else, were born into bondage . . . kept inside a prison that you cannot smell, taste, or touch. A prison for your mind," Morpheus instructs.[19]

---

[15] Deleuze, "Simulacrum and Ancient Philosophy." Here Deleuze links the simulacrum with Nietzsche's eternal return as a way of subverting Platonism.
[16] Ibid., 299.
[17] Massumi, "Realer than Real."
[18] Wachowski and Wachowski, *Matrix*.
[19] Ibid.

In such an account, the real as we know it indeed turns out to be a desert, a God-forsaken place, boring, gloomy and formidable, deathly repulsive, visualized in *The Matrix* either in the form of steel-and-glass office buildings populated with trained white-collar workers and other uniform cogs in the systemic machinery (policemen, nuns, brides), or as a scene of post-Armageddon-like destruction; an astonishing picture of a world propelled by the death drive. The poststructuralist endeavor which emptied the center of meaning understood as, and blamed for, uprooting us and leaving us unanchored, drifting thoughtlessly in indeterminacy can, and does, evoke but mere melancholy, recalling Nietzsche's last man who wills himself not to die. The Heart of the City Hotel with which, interestingly enough, the narrative begins and ends, emphasizes the fact that the very center, now occupied by the human subject, is purely a virtual construct, an illusion. Cherished hyper-individualism, that comes with the neo-Kantian turn at the end of postmodernism, can serve as one, if most insidious, version of such anthropocentrism, an obstacle to genuine revitalization. Morpheus's commandment is, therefore, a test which one must endure, "Let it all go. Fear, doubt, disbelief. Free your mind."[20] So long as one's affective attachment to subjectivity is not de-cathected—which, as the film brilliantly shows, is not at all an easy undertaking—one will remain bound to the old transcendent matrix which organizes earthly life of humans, who willingly reproduce it *ad infinitum*. This specifically Western disease,[21] which Deleuze calls "the dogma of transcendence,"[22] holds humans in check, shaping them as docile and faithful subjects, and submits them to an a priori regime of "subjectivity, signification and organism."[23] Transcendence, positing a subject, substance or being as origin, twists the order and reverses it so that images, difference and becoming appear secondary, mere products. Experience as a flux gets tamed, transcended and mediation is introduced. In effect, productive virtuality is reduced to mere reflection, representation; an illusion of the real is established—indeed, imposed on humanity from above. Consequently, as the film demonstrates, people live in a structured and regulated society, plugged to the negativistic matrix which vampirically feeds off them. In the era of late advanced capitalism, as Deleuze and Guattari emphasize, ossified systems are demoted and identities are dissolved, releasing images, information, objects that bring more new

[20] Ibid.

[21] Deleuze and Guattari, *Thousand Plateaus*, 20.

[22] Foucault, *Archeology of Knowlegde*.

[23] Deleuze and Guattari, *Thousand Plateaus*, 176.

potential and creativity than ever before.[24] Technology, a major factor in this process of deterritorialization, strikes a lethal blow at the edifice of rationalistic and humanistic civilization, exploding it and opening to the novel reality of the virtual. "Welcome to the real world,"[25] Morpheus greets the newly re-born Neo in the world of machinic artifice, celebrating his Nebuchadnezzarian awakening to the land of wonders. Deleuze states,

> There are images, things are themselves images, because images aren't in our head, in our brain. The brain is just one image among others. Images are constantly acting and reacting on each other, producing and consuming. There is no difference at all between *images, things, and motion.*[26]

What is accomplished is not a simple inversion; now the power of images (movement, becoming) is given ontological priority and epistemological validity over being. Conversely, with the simulacrum one goes beyond the distinction between being and becoming to arrive at an affirmation of eternal return as the very being of becoming. A totalizing conception of being is replaced by minoritarian becoming which is a differently named creativity; the duality of human and machine is overcome in the flow of simulacra, which only exist by dint of motion and technological inventiveness. "Everything is formed out of connections, densities, shocks, encounters, concurrences, and motions," Deleuze cites Lucretius in his *Logic of Sense.*[27] All of this happens on the plane of radical immanence that knows nothing outside itself. "In any case, simulacra are everywhere. We do not cease to be immersed in them and to be battered by them as if by waves," Deleuze continues.[28]

It is true that affirmative deterritorialization into the virtual demands great courage. As Morpheus explains: "Some people are not ready to be unplugged" from the schematic matrix.[29] As a matter of fact, the film brilliantly demonstrates that it is difficult to believe in the power of the simulacrum and its technological apparatus. Some people—indeed most of them—whose mind has already grown too old might have trouble letting go. Others, aware of such a generative and volcanic realm, whose only rest is motion, cannot bear its vibrancy and liveliness, and retreat to the old

---

[24] Ibid., see especially Chapter 13, "Apparatus of Capture."
[25] Wachowski and Wachowski, *Matrix*.
[26] Deleuze, *Pourparlers, 1972–1990*, 62 (emphasis in the original; my translation from French).
[27] Deleuze, "Simulacrum and Ancient Philosophy," 305.
[28] Ibid., 312.
[29] Wachowski and Wachowski, *Matrix*.

schematism claiming that "ignorance is bliss."[30] Others still, totally unaware of the matrix itself, do not even look for such a line of flight, a point of exit from the matrix of the real. Deleuze and Guattari add rather pessimistically, "Certain groups or people have none and never will."[31]

The war between humans and machines ended the moment that humans rethought their conceptual framework by means of which they organize and structure existence. It is only when they are able to understand that technology is no Other—bigger or smaller, good or evil—that it will dawn on them that it is precisely that which the real *is*.

> What is real? How do you define real? If you are talking about what you can feel, smell, taste, then real is simply electrical signals interpreted by your brain. World then is neural-interactive simulation, Morpheus instructs.[32]

As Deleuze and Guattari asseverate, "There is no biosphere, but everywhere the same Mechanosphere."[33] There is no room for crude binary oppositions of the natural and the artificial which are mere by-products of the generativity of the mechanistic simulacral flow. The real world, the naturalized matrix—"computer-generated dreamworld built to keep us under control"[34]—no longer has an essence save as simulated, that is as expressing the functioning of the simulacrum.[35] The Matrix emerges as a simulation; in point of fact, "The Matrix isn't real," as Morpheus explains.[36] Human beings emerge as unique assemblages of this creative, if not chaotic, flow of being. Hallward puts it succinctly: "A creative ontology is distinct from merely chaotic anarchy. Being is creation rather than chaos. Chaos creates nothing that endures."[37] This is why it is not a question of whether and how much the virtual (technological) resembles the real and of adjudicating how perfect a substitute for the real it can be; on the contrary, it is its very productivity and different mode of functioning that must be affirmed critically. Deleuze firmly states,

> Actualization of the virtual always takes place by difference, divergence or differentiation. Actualization breaks with resemblance as a process no less

---

[30] Ibid.
[31] Deleuze and Guattari, *Thousand Plateaus*, 223.
[32] Wachowski and Wachowski, *Matrix*.
[33] Deleuze and Guattari, *Thousand Plateaus*, 77.
[34] Wachowski and Wachowski, *Matrix*.
[35] Deleuze, "Simulacrum and Ancient Philosophy," 300.
[36] Wachowski and Wachowski, *Matrix*.
[37] Hallward, *Out of This World*, 29.

> than it does with identity as a principle. Actual terms never resemble the singularities they incarnate.[38]

Never can the original, the founding, the virtual matrix resemble that which it founds or produces—this is the major shift which the film succeeds in making and picturing. I, therefore, disagree with Benedikt who somewhat ironically asks, "And after all, why have cyberspace if we cannot (apparently) bend nature's rules there?"[39] Such a positing of the virtual already assumes the inevitability of ontological dualism and thus cannot reach and appreciate the critical potential that the virtual offers for our self-understanding and the notion of nature itself.

"Know thyself": Virtual Sensibility Beyond Recognition

We have arrived at a point when one can inquire about more specific or particular manners which this novel rendition of the virtual influences, or indeed, how it challenges our habituated way of being and our lifestyles. As Morpheus rightly says, "The Matrix cannot tell you who you are" owing to its radically different mode of functioning.[40] It does not provide ready-made answers and a priori definitions, does not live in assumptions; on the contrary, overthrowing traditional Western ontology and doing away with foundations, the Matrix encourages experimentation and becoming on every strata, be it philosophical, individual, social or political. Put differently, the old line of thought predicated upon the principle of originality, centrality and linearity surrenders to the real of (virtual) fluxes and forces whose only principle is that of incessant production. "Do not try to bend the spoon. It's impossible. Instead, only try to realize the truth. What truth? There *is* no spoon. Then you'll see it is not the spoon that bends, it is only yourself"—only virtual potentialities engendered on the immanent plane.[41]

This process of virtualization, however, cannot proceed exclusively by dissolving the self. In addition, if equally essentially, as Hallward puts it, "It requires the sacrifice of the most precious sacred cow of contemporary philosophy—*the other*."[42] It is exactly the evacuation of the other as a constitutive element that enables one to become infinitely creative, to unendingly become-other. Neo's preliminary encounter with the Matrix

---

[38] Deleuze, *Difference and Repetition*, 212.
[39] Benedikt, *Cyberspace*, 128.
[40] Wachowski and Wachowski, *Matrix*.
[41] Ibid.
[42] Hallward, *Out of This World*, 92.

depicts this process in which the dualism of the self and other vanishes. It obviously takes some time for Neo to accept and adjust to the new monistic vision of technological virtuality. Although first, upon transmogrification, on board of a hovercraft, Neo hesitates and disbelieves in the ultimately external status of the Other, gradually he comes to learn that it is the Other who disturbs the world, the world being but its meek and feeble product from the past. Erasure of the possibility of the emergence of anomalies (others) is enforced not by their metaphysical cannibalization but rather by absolute virtualization.

Under such circumstances, in the virtual reality of the Matrix it would be pointless to differentiate any longer between the natural and the artificial. These, as the film shows, are overcome by means of technology which institutes a singular simulacral mechanosphere. Henceforth, nature can solely be conceived of as "a process of production."[43] As Deleuze admonishes,

> Simulacrum does not replace reality, it is not an equivalent that stands for reality, but rather it appropriates reality . . . it produces reality on the new full body that replaces the earth.[44]

To be capable of such a transformation, one needs a more sophisticated and complex construal of experience, which—as the films underscores—is the threshold of the virtual, yet real, Matrix. For as long as one remains chained to the world as we know it, to its experiential epistemology, one is unable to see how to exit the dominant dualism. This in turn signals a move from experience to the conditions of experience, which goes hand in hand with Deleuze's endeavor to reunite the split stemming from Kant's account of aesthetics. Aiming at a reconciliation of the objective element of sensation (theory of sensibility as the form of possible experience) and subjective one (theory of art as a reflection of real experience), Deleuze breaks with mimesis and reflexive theory of art and—in an explicitly Nietzschean fashion—rehabilitates art and technology as virtual forces constitutive of the real. As Daniel Smith suggests, "It is only when the conditions of experience in general become the genetic conditions of *real* experience that they can be reunited with the structures of works of art."[45] The film makes this exceptionally clear in its rendering of the world as a

---

[43] Deleuze and Guattari, *Anti-Oedipus*, 3.
[44] Ibid., 228.
[45] Smith, "Deleuze's Theory of Sensation," 29.

mere simulation, "a mental projection of a digital self," an entrapment and subordination of the real to the requirements of representation.[46]

It is my contention that *The Matrix* constitutes a spectacular illustration of Deleuze's critique of the system of recognition and his re-conceptualization of experience in the form of a superior form of empiricism, i.e., transcendental empiricism. Furthermore, I claim that this is a far better-crafted framework to think the virtual and all the potential which it carries and offers in that—as Deleuze himself asserts—from the outset, this conceptual apparatus unflinchingly "directs its attack against the prestige of the negative; it deprives the negative of all its power, it refuses to the spirit of the negative the right to speak in the name of philosophy."[47]

This higher form of empiricism which is elevated to the realm of the transcendental in no way purveys a method for explaining experience; rather it is the term which captures its essentially generative and productive *modus operandi*. Deleuze criticizes Kant for establishing the transcendental instance of the subject, or perhaps subjective consciousness, on the basis of an alleged harmony of human faculties which is otherwise called "common sense." This machinery, which operates according to the underlying principle of identity, allows humans to categorize sensations as objects of recognition. Drawing on Plato again, Deleuze retrieves a distinct kind of worldly sensations, ones that force us to think, that produce thought. Deleuze calls them "encounters" or "signs" and underscores their immediate status as something that can only be felt or sensed. They, as Francis Bacon put it, "act directly on the nervous system, rather than passing through a detour of the brain."[48] Sensibility, in Deleuze's view, becomes transcendental precisely when no transcendence and mediation is allowed. It is composed not of recognizable objects, but rather sensible, if often imperceptible, forces, encounters, signs whose only way of being affirms itself in becoming, a perpetual movement of molecules connecting and disconnecting, forming and deforming, constructing and destructing, an intricate technological production, a veritable vertigo; a sensibility out of this world. The immediacy and immanence of experience permits no transcendence; the subject is a dogmatic construct of this virtual genesis which twisted it against itself, thereby proclaiming himself the ultimate condition of experience on grounds of his putative subjective identity of the self and its faculties. Deleuze states his point thus:

---

[46] Wachowski and Wachowski, *Matrix*.

[47] Deleuze, "Simulacrum and Ancient Philosophy," 315.

[48] Bacon, *Brutality of Fact*, 18.

> If this subject has no specific personal identity . . . it is because it is not only a part that is peripheral to the machine, but also a part that is itself divided into parts that correspond to the detachments from the chain and the removals from the flow brought about by the machine. Thus the subjects consumes and consummates each of the states through which it passes, and is born of each of them anew, continuously emerging from them as a part made up of parts.[49]

The human being is virtualized into a highly intricate and dynamic whole, distinguishable from other elements by means of speed and slowness. Having received a signal of connection, Morpheus transforms himself, dilutes into myriad electronic and neural flows of molecules. His identity can never be regarded as a final or completed product as is the case of the Cartesian automaton. It is above all molecular technological bodies that are in immanent interchange with their environment, a multiplicity *par excellence.* The examples of an electronic bug (machine) metamorphosing into an enlivened insect, or tracking machines swimming in the ocean of the virtual emulating octopuses, present this technologization of bodies and thus their self-renewing virtual potential for change and alteration. As such, bodies defy—or as a matter of fact, crumble—the illusions of sameness constitutive of the humanistic subject, its organic roots (organism) and its naturalized habitat (world).

> It establishes the world of nomadic distributions and crowned anarchies. Far from being a new foundation, it engulfs all foundations; it assures a universal breakdown as a joyful and positive event, as an un-founding.[50]

Virtual sensibility announces the twilight of theological givenness—"God made man in his image and resemblance"—liberating the diabolical sensible simulacrum as the immanent genetic principle of the real of the virtual, thus inaugurating posthumanity. The human being ceases to be a subject, a creator ensconced in the emptied seat of God, to become the divine art of creation itself. The agent's observation of the human species, if presented as derogatory, in fact captures the singularity of becoming of which human life is but one mode.

> You're not actually mammals. Every mammal on this planet instinctively develops a natural equilibrium with its surrounding environment. You, humans, do not. You move to an area and you multiply and multiply until every natural resource is consumed. The only way you can survive is to

[49] Deleuze and Guattari, *Anti-Oedipus*, 44.
[50] Deleuze, "Simulacrum and Ancient Philosophy," 300.

> spread to another area. There is another organism on this planet that follows the same pattern—a virus. Human beings, they are disease, the cancer of this planet, a plague.[51]

The viroid life, life as multiplication and cancerous proliferation is the life of the creative, Dionysian simulacrum and volatile sensibility which lives only by encounters.

One can already discern a certain similarity that emanates from the discussion of the simulacrum and the above reformulation of empiricism as put forth by Deleuze. These two discussions are at no rate discrepant or incommensurable. Even though the non-Platonic theory of images and the post-phenomenological rendition of experience display divergent modalities, they ultimately share a common adherence to, and insistence on, immanence. The virtual that is constructed from their inspiration does not stand for the higher version of, and hence substitute for, the real, and neither does it function as its explanatory principle. All the virtual *is* is all it does, that is effectuate, produce, construct the real. The virtual is the immanent genetic condition of the real, at no point opposed to it. As Deleuze avers,

> Pure virtuality no longer has to actualize itself since it is a strict correlative of the actual with which it forms the tightest circuit. It is not so much that one cannot assign the terms "actual" and "virtual" to distinct objects, but rather that the two are indistinguishable.[52]

One of the penultimate scenes in the film best visualizes this immanent becoming when Neo, revived to pure virtuality, merges with everything that *is* and incorporates an agent's body, shattering it into molecules. His every exhalation joins with the more profound movement that makes the walls of the corridor follow his rhythms, forming a veritable immanence.

If this is so, then the real as we know it, can by no means be conceptualized as an essence or transcendent nature, unless these concepts are themselves protected from scrutiny, or indeed virtualization, which will consequently expose their secondariness and hence their illegitimate and ungrounded status. As Deleuze emphasizes,

> The simulacrum is thus imperceptible. The image alone is sensible, which conveys quality, and which is made up of this very rapid succession, and the summation of many identical simulacra.[53]

---

[51] Wachowski and Wachowski, *Matrix*.

[52] Deleuze, "Actual and the Virtual," 114.

[53] Deleuze, "Simulacrum and Ancient Philosophy," 311.

Conclusion

At the very opening of his work on Nietzsche, Deleuze remarks that we always have the beliefs, feelings and thoughts which we deserve "given our way of being and style of life."[54] If thinking is humans' most powerful affect if, as *The Matrix* reminds us, "The body cannot live without the mind,"[55] then Deleuze urges us to make an intellectual effort and rid ourselves of all impassion and morbidity which are characteristic of the negative and revitalize thought as process, as creation. Our power to be does not affect, but *is expressive* of, our power to think and *vice versa*, as Deleuze stresses in his Spinozist spirit. Thus, the virtual cannot only be limited to the outside realm of the technological, and as such derivative of the human. It must be affirmed as equally constitutive of the human in the sense of mutual co-causality, a continuous and simultaneous *pars destruens* and *pars construens*.

It is a deeply restrictive, unfortunate and saddening manner in which to continue thinking about the Internet, the virtual space, as the Other in which, once in, we lose ourselves, condemned to a variety of negativistic postmodernist politics of "anything goes," or, in an imperialistic fashion, consider it as yet another frontier to be conquered (ordered, circumscribed, exploited), yet another space to be populated and furnished with thoughts and things which we know so well. Whilst the first evokes the all too familiar model of paranoia, the latter is but another avatar of Western narcissism, which, in the long run—as history so diligently describes—brings terror and calamity. These are exactly the perils that haunt the real of representation, which Žižek cannot help but caution us about:

> Immersion into cyberspace can intensify our bodily experience (new sensuality, new body with more organs, new sexes . . .), but it also opens up the possibility for the one who manipulates the machinery which runs the cyberspace literally to steal our own (virtual) body, depriving us of the control over it, so that one no longer relates to one's body as to "one's own."[56]

Žižek's following warning note on the dangerous liaison between the human and technology ultimately confirms his allegiance to negativity, which—although it helps understand the vagaries and vicissitudes of the postmodern world—hardly offers any exit.

[54] Deleuze, *Nietzsche and Philosophy*, 1.
[55] Wachowski and Wachowski, *Matrix*.
[56] Žižek, "Matrix."

> The commonplace is that, in cyberspace, the ability to download consciousness into a computer finally frees people from their bodies—but it also frees the machines from "their" people.[57]

As Morpheus stresses, "As long as the matrix exists, the human race will never be free."[58] Rather than mindlessly perpetuate the same old humanistic, dualistic logic juxtaposing the real (material, natural, phenomenological) world against the virtual one (dematerialized, artificial, abstracted), it is about time to think affirmatively of technology not solely as a product of human minds and bodies but as an invention of life itself, its mode and being.

I contend that Deleuze's philosophy is well-suited to help one to grasp this alternative, if not altogether new, significance and importance of the virtual. It should come as no surprise by this point that his is the immanent virtuality that allows no room for the outside, precisely because it is the only outside there ever *is*. By no means does the Real of the Virtual stand for some new all-encompassing metaphysical principle that explains everything. Rather, in allowing itself no transcendence, it radicalizes its immanent productivity and efficacy. As Jean-Luc Nancy puts it,

> Gilles Deleuze's philosophy is a virtual philosophy, in the sense which we use this word today when we speak, in a strangely indifferent way, of virtual reality or image—designating a universe entirely formed from images, and not only images as high quality illusions of the real, but rather those that leave no place for the opposition between the real and the image. The virtual world is a universe of image-effectivity.[59]

The power of the virtual is immense and its technological mode opens up new, ever-expanding fields for experimentation. It is only via affirmation that we can utilize it for our own, human, development and enhancement. As a tool of our times the virtual becomes no metaphor but the praxis for Deleuze's thinking; indeed, it is a machine that explodes the Western tradition of representation. So long as the virtual is thought of and theorized as inextricably tied up with, and thus limited to what Benedikt describes as "a globally networked, computer sustained, computer-accessed, multidimensional, artificial virtual reality,"[60] its radical and political potential is drastically circumscribed.

---

[57] Ibid.

[58] Wachowski and Wachowski, *Matrix*.

[59] Nancy, "Deleuzian Fold of Thought," 110.

[60] Benedikt, *Cyberspace*, 122.

The virtual techno-logic, effectuating a novel plane wherefrom to think and experience life, opens us up onto post-humanity and invites new manners of experimentation, which always already means that from which we have been severed—namely, creation. The film's visual and sonic expressivity testifies to this machinic functioning of the generative matrix, that is life itself: grey, misty and steely ambiance and coloring, aided by the sounds, create the viscerality and mechanicity of experience. Liberated from organizing forces of time and space, the virtual becomes the real, life realer than real as we used to know it; "a world without you, a world without rules and controls, without borders and boundaries, a world where everything is possible."[61]

Coda

To be capable of affirmation, one must renounce the old habits—habits of thought, of being, of life. Unpredictability of the future cannot impede, preclude or obliterate change. To let oneself raise an unconditional objection to the real of the established values amounts to affording oneself the right to vanish. The genuine critique, as Deleuze proves in his work on Kant,[62] cannot proceed by believing in that which it aims to criticize; a bolder act is required.

> Courage consists, however, in agreeing to flee rather than live tranquilly and hypocritically in false refuges. Values, morals, homelands, religions, and these private certitudes that our vanity and our complacency bestow generously on us, have as many deceptive sojourns as the world arranges for those who think they are standing straight and at ease, among stable things. They know nothing of this immense flight that transports them, ignorant of themselves, in the monotonous buzzing of their ever quickening steps that lead them impersonally in a great immobile movement.[63]

The flight into the virtual . . .
May we never be afraid to follow the white rabbit!

Works Cited

Bacon, Francis. *The Brutality of Fact: Interviews with David Sylvester*. London: Thames and Hudson, 1975.

---

[61] Wachowski and Wachowski, *Matrix*.
[62] Deleuze, *Nietzsche and Philosophy*, 90; *Kant's Critical Philosophy*.
[63] Deleuze and Guattari, *Anti-Oedipus*, 374–75.

Benedikt, Michael. *Cyberspace: First Steps.* Cambridge, MA: MIT Press, 1991.

Deleuze, Gilles. "The Actual and the Virtual." In *Dialogues II*, by Gilles Deleuze and Claire Parnet, 112–15. London and New York: Continuum, 2006.

—. *Difference and Repetition.* New York: Athlone Press, 1994.

—. *Kant's Critical Philosophy: The Doctrine of the Faculties.* London: Athlone Press; Minneapolis: University of Minnesota Press, 1984.

—. *Nietzsche and Philosophy.* New York: Columbia University Press, 1983.

—. *Pourparlers, 1972–1990.* Paris: Les Editions de Minuit, 2003.

—. "The Simulacrum and Ancient Philosophy." In *The Logic of Sense*, 291–316. London: Continuum, 2004.

Deleuze, Gilles, and Félix Guattari. *Anti-Oedipus: Capitalism and Schizophrenia.* London: Continuum, 2004.

—. *A Thousand Plateaus: Capitalism and Schizophrenia.* London: Continuum, 2004.

Dorter, Kenneth. "The Ion: Plato's Characterization of Art." *The Journal of Aesthetics and Art Criticism* 32, no. 1 (1973): 65–78.

Foucault, Michel. *The Archeology of Knowlegde and the Discourse on Knowledge.* New York: Pantheon, 1972.

Gibson, William. *Neuromancer*. New York: Ace Books, 1984.

Hallward, Peter. *Out of This World: Deleuze and the Philosophy of Creation.* London: Verso, 2006.

Haraway, Donna. *Simians, Cyborgs and Women: The Reinvention of Nature*. New York: Routledge, 1991.

Lawlor, Leonard. "The Beginnings of Thought: The Fundamental Experience in Derrida and Deleuze." In *Between Deleuze and Derrida*, edited by Paul Patton and John Protevi, 67–83. New York: Continuum, 2003.

Massumi, Brian. "Realer than Real: The Simulacrum according to Deleuze and Guattari." *Copyright*, no. 1 (1987).

Nancy, Jean-Luc. "The Deleuzian Fold of Thought." In Patton, *Deleuze: A Critical Reader*, 107–13.

Patton, Paul, ed. *Deleuze: A Critical Reader.* Oxford: Blackwell Publications, 1996.

Smith, Daniel W. "Deleuze's Theory of Sensation: Overcoming Kantian Duality." In Patton, *Deleuze: A Critical Reader*, 29–56.

Wachowski, Andy, and Larry Wachowski. *The Matrix*. DVD. Directed by Andy Wachowski and Larry Wachowski. Los Angeles, CA: Warner Bross Home Video, 1999.

Žižek, Slavoj. "The Matrix: The Truth of the Exaggerations." http://www.lacan .com/matrix.html (accessed March 24, 2009).

From the Visible to the Visual: Negotiating the Technological in American Cinematic SF

Dorota Golańska

Visual culture should be seen as an exceptional field of postmodern contestation. On the one hand, as a medium for spreading the dominant ideology it popularizes the defined and naturalized cultural and political meanings. On the other hand, however, visual culture supplies us with the space for resistance against the dominant meanings and myths and as such can be seen as a means of political struggle against universal paradigms of knowledge and power. Serving as a forum for interactions, tensions, and contestations, visual culture starts to contemplate itself, plays with well-known conventions and undermines the most disseminated paradigms.

Since visual studies are centered on a broadly understood concept of the visual, the focus of this paper is on the issue of visualization as a theme debated upon within the spaces of visual culture. The analysis is confined to the contemporary American cinematic science-fiction, whose field of interest has recently shifted towards not only biomedical and genetic themes but also towards the issues connected to linguistics and epistemologies of knowledge. The analysis is even more interesting if we take into account the fact that the genre developed as a result of unprecedented enchantment with new technologies, inspired by the dream about achieving everything that was possible and the belief that technology would equip the human being with entirely new possibilities. Contemporary science fiction, however, seems to undermine and contest the well-established paradigms, which have previously supplied the genre with an important conceptual basis. My aim is to point out that popular culture creates a space for discussing the issues of postmodern techniques of visualization and control in the context of diversity and difference and with an aim to overcome the Western dualistic logic of the subject and object of knowledge. For this reason, I also consider the concept of the human body especially meaningful: together with the development of new technologies and medical sciences, the body has started to be seen as an object of increased contestation. The human body is marked with difference and as such it is located within the relations of power, where ideologies and dominant meanings are projected. Undermining the paradigm of seeing, cinema—which itself relies on the visualization techniques—contributes to a debate within the field of visual studies, questioning most popular practices and ideologies, transcending alleged boundaries, and overcoming paradigms permeating both science and popular culture.

My argumentation, which combines philosophical reflections on medicine and film, relies on the assumption that between science (itself a realm of culture) and culture (including popular culture) there exist a network of intense interactions and inter/intra-connections. As has been pointed out by, among others, Jon Turney[1] and Claudia Springer,[2] not only is culture very often inspired by science and vice versa, but both create a space for the production of ideologized meanings as well as a platform where the dominant meanings are negotiated. The increased hybridization of contemporary life, as pointed out by, for instance, Bruno Latour[3] and Donna Haraway,[4] requires a thorough analysis which is not limited by disciplinary boundaries. Since the disciplines are deeply intertwined, an analysis of popular culture in the context of scientific paradigms dominating the modern medicine is well-grounded, taking into consideration the fact that both rely on and cannot exist without technologies and practices related to the issue of visualization. The reason for using science fiction to look at these attitudes is the genre's position at the unique intersection of science and technology, mass media, popular culture, literature, and secular ritual.

Visualization and the Modern Episteme

Donna Haraway claims that technologies of visualization can be compared to the practice of hunting with a camera and evokes a very predatory nature of photographic consciousness. As she claims, contemporarily, in the era of globalization and due to the immense impact of technologies, we have to deal with a complete disappearance of an unmediated experience.[5] Social life is marked with technological mediation which entails a thorough redefinition of the alleged schemes of

---

[1] Turney, *Frankenstein's Footsteps*.
[2] Springer, "Psycho-cybernetics in Films."
[3] Latour, *Science in Action* and *We Have Never Been Modern*.
[4] Haraway, "Cyborg Manifesto."
[5] Ibid.

thinking which must be adjusted to the new social reality. With the help of contemporarily available technologies of visualization, information and communication, cultural phenomena which nowadays take place both in medicine and cyberspace can be seen as a perfect analogy of poststrucuralist philosophy, where meaning and object are separated, or as a metaphor for a socially constructed reality. The value of visual representation, on which contemporary culture is focused, is manifested in the permanent necessity of being confronted with images and—simultaneously—being the object of the omnipresent gaze. The priority of vision has supplied the paradigm of modern science with an important basis and has evolved into a guarantee of knowledge. In the context of contemporary culture, seeing is usually interpreted as knowing. This paradigmatic shift implies an increased transparency of natural and social processes—the possibility of seeing, often aided with highly sophisticated technological equipment, is a *sine qua non* of knowing, understanding, controlling, and manipulating. As knowledge entails purely material consequences, the active role of the observers in the production of scientific facts needs to be acknowledged.[6] The complex and ambivalent role of technology cannot be overestimated. The binarized framework of Western science seems to be incapable of processing the critical potential of technology. I want to stress that the new conceptualization of the technological (not as a copy but as an immanence of the real) plus the dissolution of the alleged boundaries between the natural and the technological, which parallels the dualism between the Self and the Other, all signal the demise of the rationalistic legacy and humanist bias inscribed in it. Haraway[7] claims that the anthropocentric modern universe is built upon the dualistic system, which introduces such oppositions as real/artificial, natural/cultural, masculine/feminine, young/old, analytical/emotional, past/present, alive/dead, and human/technological. It is important to keep in mind that also the poststructuralist philosophical endeavors to finally divorce the copy from the original, where the latter eventually falls out, perpetuates the well-established Western binarized logic and does not let us abandon the negative thinking. Consequently, we remain locked up in the dualistic framework of a copy and an original. Transgression of these dualistic barriers constitutes—according to the posthuman view—an important conceptual point of departure for objective judgments.

---

[6] See Barad, "Posthumanist Performativity, 801–31; or Rouse, "Barad's Feminist Naturalism," 142–61.

[7] Haraway, "Cyborg Manifesto."

It is often argued that the fast development of new technologies has significantly contributed to the pluralization and democratization of available perspectives and subjectivities. Contemporary times have witnessed an awesome alliance of *techne* and *episteme*. Posthumanism celebrates the final dissolution of the liberal humanist subject and favors the existence of diverse and often contradictory subjectivities within one and the same person. Contrary to this view (especially popular among the Internet/cyberspace enthusiasts), I would like to argue that the new achievements in science, above all these connected to the enhancement of the methodological and technological scientific instrumentarium and equipment, often contribute to the development of transcendental models and schemes, which trigger the well-established ideological divisions and perpetuate the modern way of thinking in accordance with the alleged categorizations, so well-established within the Western philosophical framework. Traditionally, the task of *techne* was relegated to the auxiliary application, which had to be anteceded by the original work of understanding. Thus, the technological other, doomed to subservience, played a constitutive role in sustaining the system of Western dualisms. Consequently, "otherness" has been loaded with pejorative meaning. The technological "Other" as a cinematic figure, used and abused in American science fiction, has for many years served to exemplify the American preoccupation with either negatively or (more recently) positively conceptualized difference. The process of the coexistence of the real and the technological proved to abound in varied representation of the latter (as opposed to the former), ranging from the deification of technologies to their extreme brutalization, accompanied with radically polarized feelings. As Monica Casper points out, the concept of human agency draws on a dichotomous ontology, which assumes the opposition between the human and the non-human. This kind of view stems from political practices referring to the abovementioned heterogeneous division.[8] In this paper, I use the figure in a more metaphorical sense—artificiality is thus understood very broadly, including social constructedness and ideological interpretation of all human and nonhuman processes. Scientific visualization technologies—which are the major focus of this analysis—supply us with an extension of vision, a sense that is seen as a universal way of experiencing the external world. Contemporary culture, a heir to the positivist epistemological model, assumes a total transparency. If seeing is understood as the *sine qua non* of knowing, the mental act of seeing is considered to be the universal model of acquiring knowledge.

---

[8] Casper, "Reframing and Grounding," 839–56.

The consequences that the further development of visualization techniques entails are very important in the context of the reciprocal relation between knowledge and power already pointed out by Foucault.[9] Although cybernetic posthumanism heralds the plurality of available subjectivities and intense contradictions within the same subject, I would like to argue that the negligence of the corporeal dimension of human existence can result in dramatic consequences not only for the physical existence of the human being as such, but for broadly understood science. In terms of the modern epistemology of knowledge, which nowadays depends on technological enhancements and relies on well-established models dating back to the scientific revolution and the philosophy of the Enlightenment, diversity of perspectives is not yet widely acknowledged.

Technology and American Culture

The history of the West since the discovery of America can be seen as a whole chain of often dramatic upheavals, predicated on technology advancing with geometrical acceleration, and affecting all aspects of human experience. In my opinion, analysis of science fiction movies is important since the genre is an expression of the dream about nature and the control of nature. It is a peculiarly American dream, because of the nation's special relation to both technology and the myth of the Earthy Paradise in the New World, as argued already by Leo Marx.[10] What makes science fiction important in the U.S. culture, however, is not only that its tropes contain complex ideological and mythical views of nature. As Sharona Ben-Tov argues, science fiction is a mass dream with an imperative: a dream upon which, as a nation, Americans act.[11] In science fiction's rich, synthetic language of metaphor and myth we can trace the vital connections between such diverse elements as major scientific projects, the philosophical roots of Western science and technology, and American cultural ideals. Science fiction combines frontier mythology with myths about collective technological power. It does so in an effort to define an American (white) identity. There exists a vital connection between the frontiersman (and I use the gendered term advisedly) and the scientist. On the one hand, both in science and science fiction the solitary violence of the frontiersman is civilized into a collective technological power over the passive and inanimate world of nature. On the other hand,

[9] Foucault, *Birth of the Clinic*.
[10] Marx, *Machine in the Garden*.
[11] Ben-Tov, *Artificial Paradise*, 2.

the technological system accompanied with the epistemological issues well-established in the Western culture is naturalized so that it becomes second nature for Americans/Westerners. These are the questions that place both science and science fiction in the context of popular frontier mythology. Since the beginning of white settlement, the myths of the frontier have affected American attitudes by exalting violence on nature into a purifying, redemptive, and even generative force. Science fiction reiterates, but also changes, the idealized violence of frontier mythology. Frontier mythology is about the desire of limitlessness, and this is the essential theme of science fiction's American meditation. When science is rhetorically constructed as an exercise in domination, what it searches for and finds in the study of nature will certainly be, among other things, the power to dominate. The positivist epistemological model was created as a tool to fulfill these goals; however, recently it has been put into question and new, alternative approaches to science have begun to challenge the mainstream.

In this paper I argue that contemporary American science fiction is increasingly preoccupied with issues of knowledge and power, and that it focuses on the epistemological problem of technologically enhanced vision. At the same time, the technological "otherness" remains one of the most popular themes of the genre, although its use has changed. This conceptual shift is meaningful, as it not only allows us to think critically of our technology-dominated times, but also demonstrates that the teratological imagination of contemporary culture can help us reconceptualize the difference understood in terms other than technological. The analysis of two American science-fiction movies, *Minority Report* (2002) and *AI* (2001), both recent and both by Steven Spielberg, serves to exemplify the conceptual revaluation of the genre.

The American frontier hero's relationship with nature and the dominant model of Western science were given a philosophical basis by Ralph Waldo Emerson. In his essay "Nature" Emerson imagines himself as a transparent eyeball, a clear medium into which impressions of the external world enter. But—as Ben-Tov argues—"his [Emerson's] gazing does not stop at the trees in his immediate line of vision; it expands outward. 'Standing on the bare ground—my head . . . uplifted into infinite space . . . the currents of the Universal Being circulate through me,' till the spirit of the observer expands to include the infinite universe."[12] Further, Emerson argues that "[t]he self transcends its mortal limits by taking total

[12] Ibid., 133.

possession of an actual world."[13] Evelyn Fox Keller stresses the outstanding value which vision has for Western knowledge; it is seen as the most important sense and is not comparable to any other. Keller claims that this exceptional status of vision makes the division between the object and the subject of research possible. The typical scientific attitude, which the positivist model favors, assumes distanced and detached observation. Only such a "neutral" perspective is perceived as objective since it does not imply any particularity of either the subject or object of knowledge and denies any interaction between the two. This ideal perspective is only possible to achieve for very well-defined subjects of knowledge, which for Keller means that the whole body of knowledge has to be dependent on the system of dualisms. Since the mind (the right manufacturer of knowledge) is identified with masculinity (and the opposition between the subject and object of knowledge is created analogically to the opposition of mind and body), the rational subject is able to establish a distance between himself and the object of knowledge.[14]

The best and most popular example is the famous picture of the planet Earth taken by NASA, which has already drawn the attention of Diamond and Orenstein[15] or Barry and Wernik.[16] The globe as seen from space is there objectified with the gaze of an eternally distanced observer. This seemingly innocent visualization serves as a perfect metaphor for the dominant scientific model assuming dehistoricized, "objective" observation, aided by increasingly enhanced technology. The panoptical view is defined by a specific perspective articulated with the positivist epistemology. Its attributes include the decontextualized gaze dominating the picture and powerful enough to control its content. At the same time, however, this ideologically constructed camera gaze is hidden and passes as the most natural of all perspectives. It covers the ideologized apparatus involved in the production of every gaze, in this case made possible by the technological achievements in media and communication techniques. Such a construction of the camera view creates the illusion of the possibility of acting from the omnipotent and omniscient epistemological perspective, typical for Western culture, which Donna Haraway has once called a "god-trick."[17]

The logic of contemporary culture—as problematized by Jean Baudrillard in *L'échange symbolique et le mort* (*Symbolic Exchange and*

---

[13] Ibid.

[14] Keller and Grontowski, "Mind's Eye," 207.

[15] Diamond and Orenstein, *Reweaving the World.*

[16] Berry and Wernik, *Shadow of Spirit.*

[17] Haraway, "Promises of Monsters."

Death)[18]—relies on the fact that the representation comes before that which is represented. The represented object in hyperreality becomes the object of visual consumption to such an extent that its presence is no longer necessary. The dualism of an original and a copy, however, only seemingly collapses. In fact, it becomes even more extrapolated than before, since the distance between the two is only broadened to infinity. In the public imaginary, the human material body has recently become such an object, which—on account of excessive visual representation—vanishes and becomes a pure Baudrillardian simulacrum. This way of thinking is a result of modern philosophy, for which the Cartesian *cogito* paved the way. Philosophical rationalism has created an unbridgeable gap between the human mind and body, privileging the former and marginalizing the latter. This dualism constitutes a conceptual basis for all other modern dualisms, such as, for example, culture/nature, human/nonhuman, masculine/feminine, same/other, original/copy.

Currently, we need to deal with the apparent disappearance of the human body. On the one hand, this phenomenon can be observed due to the fast development of information and communication technologies that seem to promise the final liberation from the confines of corporeal physicality. On the other hand, the process is triggered by the modern visualization techniques widely applied in medicine, which mediate the human experience of their own body and transform the latter into a set of mechanical processes represented in the form of data which can be scientifically modified and rearranged according to the human will. Visualization techniques contemporarily available in medicine, combined with new media techniques, hugely contribute to the fact that the inside of the human body is increasingly exposed to human vision—it is present in the documentary, arts, media-relayed surgical treatments, TV news, digital visualizations, etc. Consequently, the common perception of the human body has changed significantly. Modern medicine combined with media contribute to the fulfillment of the desire to have more access to human physiology and biology, which results in the transparence of the human body and in making it an object of control. The world of medicine cannot be thus limited to the scientific field but it gets increasingly intertwined with the world of culture; their reciprocal influence should not be overestimated. For these reasons, the concept of the human body is an object of significant revaluation, since the visual and technological mediation of the experience of the physical body contributes to the deepening of the aforementioned dualism which results in the

[18] Baudrillard, *L'échange symbolique et le mort.*

"disembodiment" of the medical practices while making their detailed and close observation possible. The ideal of the transparent body is founded on the radical separation of the real body and the techniques of studying it; the mediated image is always just an image, and the process of observing the human body is currently limited to observing its visualizations and analyzing the information seen as the most important substance of the living organism.

As pointed out by Michel Foucault, human bodies have recently become a space where human eyes and organs meet, which results in the "externalization of the inside."[19] Not only does this gaze dominate in medicine but it has spread to other realms of human activity. To the same extent as information and communication technologies, modern medicine contributes to the process of reducing it to a set of mechanical processes, which is in keeping with the ideal of transparency, deeply rooted in Western thought. Ann Balsamo argues that the whole set of visualization techniques triggers the fragmentation of the body and its reduction to particular organs, fluids, and genetic codes, which boosts the idea of conscious self-regulation and results in the transformation of the body into an object of intense surveillance and control.[20] José van Dijck claims that human bodies—mediated with either visualization technologies or media techniques—have been identified with the idea of transparency. The argument is as follows:

> Mediated bodies are intricately interlined with the ideal of transparency. Historically, this ideal reflected notions of rationality and scientific progress, more recently, transparency has come to connote perfectibility, modifiability, and control over human physiology. The ideal of transparency is not simply pushed and promoted by medical science. The transparent body is a complex product of our culture—a culture that capitalizes on perfectibility and malleability. In our contemporary world, the interrelations between medicine, media, and technology are all but perspicuous.[21]

The overlapping issues of vision, visualization, transparency, power, and control are currently discussed within the space of American cinematic science-fiction. In what follows, I would like to point out several manifestations of the genre's preoccupation with problems connected to science and epistemology. The appearance of new themes provides evidence for the conceptual shift currently experienced by both

---

[19] Foucault, *Birth of the Clinic*, xiv.

[20] Balsamo, *Technologies of the Gendered Body*.

[21] Dijck, *Transparent Body*.

science and popular culture. It is especially meaningful in the case of the American context, long dominated by the myth of the frontier which has currently moved not only to cyberspace but also to the inner space of the human body. Situated in the context of ferocious discussion on marginality, multiculturalism, and dominant ideology, the new fields in which American SF is interested can contribute to a thorough revaluation of the notion of "difference" (constructed on a set of Western binarisms) and its new reconceptualization in positive rather than negative terms.

Negotiating the Technological

In the opening scenes of Steven Spielberg's *AI* the viewers are familiarized with the scientific views that are important for the further narrative of the movie. These are presented in the speech made by Professor Hobby, in charge of designing androids. Firstly, Hobby introduces a clear-cut boundary between the simulation of sensuality (a function of the robot-lovers manufactured by Cybertronics) and real love, defined as the feeling that children have for their parents. Secondly, as stressed by the professor, the feeling of love is an important point of departure for the development of all other feelings which would constitute a basis for consciousness, understood as the ability to produce systems of meanings, symbols, metaphors, and dreams (i.e. the ability to represent). This consciousness would be the same as the human one. When making this claim, Hobby differentiates between the human and non-human world, and in so doing he formulates the following (third) hypothesis, well-grounded in scientific research. He implicitly denies the fact that previously manufactured robots had abilities to develop consciousness and feelings since their destinies were limited to well-defined purposes and functions. When Gigolo Joe—a robot-lover—talks to David and tries to convince him to give in to his objectives, the viewer may assume that he confirms what Hobby has stated at the beginning of the narrative. Hobby claims that humanity is scientifically defined by the ability to develop feelings and, above all, the ability to love. For this reason, regular robots are put in opposition to human beings. However, new achievements in science have provided the human being with the knowledge and abilities necessary to create an entirely new robot, revolutionary in terms of its ability to love. Stating these claims, Hobby makes explicit the main assumption of the dominant paradigm of science. It may be inferred that the narrative will develop in accordance with the rules sketched at the beginning. Nevertheless, there are some premises that undermine or debunk the professor's claims. The difference between the natural and artificial being is made increasingly more problematic and complex, as the paradigms of science are put into question.

As implied in the movie, David—the first robot who has abilities to develop human feelings and consciousness—is not in fact the first at all. Taking into account the way the other robots act—sometimes contrary to what they tend to say about themselves (and thus contrary to how they represent themselves)—we can point to significant discrepancies between the scientific theory and everyday praxis. Although Gigolo Joe tries to persuade David to return to his originally designed functions, he himself is keen to abandon his destiny once his existence is put in jeopardy. This forces the viewer to consider the way in which ideology operates in the society and lays bare the constitutive power and tremendous impact the dominant discourse has on people's identities, as well as the way people are inclined to think about themselves. The ambivalent and often contradictory construction of the android figure on the one hand, plus the discussable nature of the human abilities to feel and love on the other, result in the blurring of the clearly defined boundaries to which Hobby's speech refer.

In *AI*, the priority of vision is articulated through the preoccupation with technological achievements that make a thorough imitation of the human organism possible. Sheila—a female android serving as a model for Hobby's presentation—is asked to show her mechanical mind hidden behind the mask of artificial skin that covers the rationally designed mechanism. Although forced to perform this exhibitionist act (also in the literal sense), Sheila will not abandon the function she has been given. This meaningful scene is undoubtedly one of those crucial for understanding the movie. The presentation of the artificially created brain exemplifies the preoccupation of the Western culture with the inclination to make nature disclose her secrets. The study of nature is treated as a means to achieve the total transparency of biological and physical processes, as well as the transparency of social life. The same is suggested by another scene from *AI*, when little David is taken to the clinic to clean up his insides after the meal he should not have eaten. The operation scene is shot in such a way that it resembles common medical treatments relying on complex technological equipment. Aided with medical tools, human vision is taken to a place previously hidden from the common gaze. As a result, the inside of the human body tends to be perceived as a model thoroughly known by scientists and made visible to everyone who wants to see it. The proliferation of images depicting the inside of the physical body results in a situation when people start to think about their bodies as an external object of minuscule, detailed, and detached study.

The facts presented by Hobby are only valid in "laboratory" conditions, to use a term coined by Latour.[22] They are fictions operating in the social context, a basis for material practices and as such constructing rather than reflecting reality. As suggested in the movie, the *mechas* that preexisted David already possessed the abilities defined as human and in many respects behaved more humanly than the humans themselves. Contrary to Hobby's claims, humanoid robots from Spielberg's movie cannot be treated as usual artifacts. Their psychic construction makes them comparable to human beings and this confrontation serves to expose the social construction of the latter, their reliance on cultural clichés and attachment to schematic behaviors. The ambivalence of the division between humans and nonhumans in *AI* can be seen clearly in the confrontation between Martin (Monica and Henry's biological son) and David (the mechanical child adopted by the couple). Although Martin tends to treat David as a supertoy (and compares him to Teddy), there are more similarities than differences between the two. Even if they are classified as belonging to two distinct ontological categories, in many respects Martin resembles the classical cyborg, who relies on technological enhancement and scientific achievements in order to function. His life is not continuous, interrupted by a period of artificial coma in a cryogenic clinic where his vital forces are sustained with complex technological equipment. The case of mechanical Teddy, turned off and put away in the closet for the time of the boy's sickness, parallels the non-continuity of Martin's life. When recovering from the cryogenic coma, Martin's body is sustained with drips, oxygen masks, and other medical apparatuses, while his movements are aided by a wheelchair and braces. Such a construction of the movie's narrative seems to imply that the human being has become increasingly dependent upon technologies which not only facilitate their life, but very often are a *sine qua non* of their existence. The ever-present distinction between the two boys seems to be ungrounded and artificially constructed.

Contrary to expectations, the movie portrays the natural humans as more mechanical than the artificial ones, and in this way identifies the former with robots/machines. This ambivalence is at the source of the *orgas*' hostility towards the *mechas*, since robots challenge the humanity of the human beings and the fear of humanoids as presented by Spielberg is in fact connected to the fear of disclosure. The meaningful figures of robots make explicit the mechanical behavior of people determined with imperceptible social clichés which structure human experiences. The

---

[22] Latour, *Science in Action*.

cultural masks rely on gestures. How should we otherwise explain the fact of the make-up adjustment performed by humanoid Sheila, whose face has been created as impeccable? The gesture must be thus understood as the enactment of a cultural ritual, especially in the context of the scene which follows, when Monica repeats the same act while on her way to the cryogenic clinic. The same kind of ritual is replayed several times at different moments of everyday life, and David, non-acquainted with cultural schemes, deconstructs them and highlights their artificiality—the social constructedness of rituals that otherwise pass for natural. When eating, David mimics people's gestures, although in his case the activities do not make any sense. The family dinner scene should be interpreted as an apogee of this mystification. Suddenly David bursts into terrifying laughter not at all commensurate with the situation, but soon afterwards both Monica and Henry are laughing in a way no less artificial or disturbing, which should be thought of as a mechanical and necessary response to David's behavior. This reaction is culturally determined and unwittingly released. It is not a coincidence that this event finally convinces Monica to make use of the imprinting procedure which will make David love his new parents.

In Spielberg's movie human beings are portrayed as artificial in the social meaning of this term, since their everyday life—seemingly the most natural activities—are mechanically ritualized and loaded with symbolical meanings. *AI* seems to be playing with the view that people are more mechanical and artificial in their behaviors and gestures than they usually consider themselves to be. The artificial/technological organism—a machine—discloses this human trait and blurs the boundaries between the two seemingly distanced and distinct ontological zones. In *AI* the fear of disclosure is articulated through the narrative's omnipresent preoccupation with the opposition between the artificial and the natural, and results in a negation of the technological beings. Although deconstructed in the movie, such a composition of the narrative parallels the opposition between "same" and "other," where the latter is charged with a pejorative connotation. It means that the seemingly universal and objective paradigm of science presented in the opening sequence of the movie is put into question. As the narrative proceeds, Hobby's hypotheses are undermined and debunked, one by one. It is implicitly suggested that science is not immune to the operation of the dominant ideologies and thus that scientific facts are socially constructed. While the everyday rituals and social clichés in which the Swintons participate are exposed and deconstructed, it is impossible to resist the idea that the paradigm of science is also a result of the intertwined and interested conventions and as such is completely devoid of innocence. As implied in the movie, this paradigm—though presented as the only possible option—is just one among many available perspectives. The dominant view passes for the neutral, impartial, and universal one; yet this is a result of a very specific point of view, though its situatedness is not acknowledged. The analysis of the movie's narrative encourages a reflection on the issue of scientific objectivity (understood as value-neutrality achieved through detached and distanced observation): should this paradigm be seen as resulting from mechanical and culturally determined generalizations and omissions? Does it include minority perspectives? Is its universality a result of ideology apparatuses? Spielberg seems to purposely question the value of scientific knowledge thus constructed.

"We see what they see," says the technician whose task is to look after the precognitives in the *Minority Report.* During the six years for which Precrime has operated, Anderton has been in charge of reading and interpreting the precognitives' visions. In order to accomplish his tasks, he makes use of light-gloves and big screens on which the images are projected—he is able to zoom in on selected items, omit the less important ones, and add information from different databases. In this context, the precogs serve as authors (or mediums for information) who deliver texts (in the form of images) which are then read and interpreted by Anderton. He is responsible for arranging the pictures and excerpts into a coherent entity and thus creating a story which is going to happen, identifying the main actors, finding the place and, finally, preventing the crime. Although apparently efficient and flawless, the system has been built upon an unstable basis as it mostly relies on Anderton's interpretation and the seemingly consistent yet fragmentary visions of the three precogs. The visual evidence materialized in the form of images is extremely dynamic, the pictures flash and flicker on the holosphere with such speed that interpretation hardly seems possible. When slowed down, magnified and rearranged, the scenes and sequences are organized into a coherent cause-and-effect relationship of events which results in a logical story. The whole process is dependent upon the interpretation of the allegedly objective and neutral visual evidence, without taking into account (as in the case of the positivist model) the fact that interpretation is always effectuated from a very specific epistemic position. The same can be said about the subject of interpretation who is always located in a particular psycho-social context (and, as is revealed further in the narrative, the context of the protagonist is not at all neutral and significantly determines his point of view).

The scenes in which Anderton is reading the precogs' visions projected on the transparent holospheric screen are shot as if the director wanted to imply that the protagonist is not an external observer of the events being analyzed, but rather is situated at the centre of action. This means that his perspective is subjective. The course of action becomes even more complex when Anderton seems to wrongly interpret the sequence of events originated from the precognitives' minds and connected to the crime of which he himself is the perpetrator. As alluded to in the vision, Anderton kills a man, Leo Crow: he points a gun at him, the gun fires, Crow is hit with a bullet, and, wounded, falls out of the window. The whole incident is observed by a third man, who wears sunglasses and watches the other two through the window. Yet, when the predicted situation starts to unfold, some previously unrecognized or unacknowledged circumstances radically change Anderton's point of view: the man with the sunglasses is a billboard picture; the scene of the crime is arranged in such a way as to provoke Anderton's aggression; a former criminal, Crow is not responsible for Sean's kidnapping; finally, Anderton resists the temptation to kill Crow—the latter's death results from a clash between the two and nobody knows who actually pulls the trigger. The visual evidence reconstructed from the precogs' visions and deemed sufficient to arrest Anderton significantly differs from the real events, which questions all previous interpretations of all precedent visions which supplied the Precrime with arguments for the detention of "future" criminals.

Although all technological instruments which transmit data (most often materialized in pictures) are presented in the form of transparent screens, it is explicitly suggested that the visual evidence is not at all transparent and the technology necessary to produce it, combined with human interpretation, encourages endless manipulations and distortions. The fact that the so-called minority reports are rejected in the process of generalizing evidence evokes the contemporary epistemological debate on the issue of "false" objectivity.[23] Hence, it may be argued that, potentially, each vision can be considered a minority report, especially on account of the fact that none of them are fulfilled: arriving at the scene of the crime and arresting the alleged villain, the Precrime changes the course of the forecasted story. The majority vision (i.e. the dominant one) is acknowledged as the right one, and the visual evidence is a result of a dramatic generalization which relies on detecting the dominant schemes,

---

[23] See Harding, *Science Question in Feminism*; Haraway, "Situated Knowledges"; and Grosz, "Bodies and Knowledges."

as well as on a simultaneous marginalization of the minority options. In the movie, this argumentation is articulated in the figure of the precog Agatha who—while being the key element of the whole system—knows that the future that she sees in her visions is just one of the available alternatives. She sees the future and nevertheless understands how dangerous it would be to trust it. There is a scene when Anderton and Agatha—symbolically putting their heads on each other's shoulders—form a two-face figure watching simultaneously in two different directions, as if they were trying to see two very distinct futures or representing two opposite points of view. This meaningful figure also points to the plurality of possible perspectives even within the same person, which alludes to the concept of the posthumanist amalgamic subject, often inherently divided and internally conflicted. The assumed transparency of the social processes which are the focus of the Department of Precrime is significantly undermined and put into question.

Minority Report is centered on the issue of seeing and visualization as a means of obtaining knowledge and exerting social and political control. The movie features a pervasive use of imagery, semantics, and intertextual references related to vision. This is suggested early in the opening scene when the viewer is confronted with a set of unfamiliar, incoherent, and blurred images. In voice-over, we can hear the meaningful phrase "I can't see without them" and finally the gaze of the camera zeros in on the iris of Agatha's eyeball. Seeing becomes also the very identification of the subject (who passes through constant laser eye scans to determine their identity, intentions, and whereabouts). A series of inversions introduces the ironies involved in the process of debasing these very identities. The importance of vision is alluded to at other moments of the movie, either as part of the narrative—when Anderton needs to have a pair of new eyeballs implanted in order to hide and find the information he needs—or when the metaphors of vision are used in the script ("We have eyes on them," "Can you see?"). Anderton indulges in trance sessions induced by a drug called "clarity" that allows him to access the nostalgic world of his memory. Along with the drug, his dealer—ironically, an eyeless criminal—also passes on a piece of Biblical wisdom: in the kingdom of the blind, the one-eyed man is king. The transparency of visual evidence is paralleled with the transparency of all mental and social processes since the visual analysis of the motives for potentially harmful behaviors serves as evidence for penitentiary policy towards the assumed villains. The increased development of invigilation and surveillance technologies opens up new possibilities of manipulation, imposture, or camouflage. The omnipresence of cameras seems to make any interpretation redundant, assuming that the information can materialize in the form of visual evidence, which is considered impossible to undermine. Contrary to expectations, the multiplication of monitoring devices encourages self-control, which means that further surveillance is no longer necessary.

As suggested in my analysis of the two movies, the visual evidence is immune neither to deliberate nor non-deliberate distortions. Every interpretation—as pointed out by feminist and post-colonial critiques—is situated; however, the masculine Western perspective assumes universality of knowledge thus produced. It is achieved through negating the particularity of the position from which the knowledge originates. Culturally constructed as objective, the paradigm of vision has recently become an object of a broad critical debate which aims at disclosing its pretensions to neutrality.

Conclusions

The excessive focus on the process of seeing as a means of achieving information and knowledge is a natural result of the hugely ideologized Western models relying on certain dualisms and further perpetuated by poststructuralist logic. The artificial division of the world into two distinct spheres of culture and nature implies further binary oppositions, also manifested in the dichotomization of the scientific process, where we need to deal with the active subject and passive object of observation. This alleged gap is only possible in the case of distanced and decontextualized gaze, which triggers the dramatic opposition between activity and passivity. In the era of simulation—to use Baudrillard's term—it is vital to pay attention to the issue of artificiality, whose definition must be broadened to include the social constructedness of the human being. As Baudrillard claims, living in the postmodern world is comparable to the state of pure presence, where everything seems to be transparent, visible, and exposed.[24] In these circumstances, human bodies have also become the object of unemotional and distanced gaze—objects deprived of any internal dimensions, desires, or purposes. Everything is "visible, too-visible, more-visible-than-visible." In this sense, the human body is understood as the final cultural metaphor for power and domination and is regarded as the final scientific frontier.[25]

Together with the appearance of a broad range of scientific approaches underlining the network of intense inter- and intra-actions of different

[24] Baudrillard, "Ecstasy of Communication."

[25] Crawford, "Cultural Account"; Bordo, *Unbearable Weight.*

ontological spheres (triggered by the unprecedented development of new technologies),[26] such as, for instance, extreme social constructivism, the positivist model of science has begun to experience a thorough crisis, while still being the dominant one among many academic communities.[27] For example, Karen Barad stresses the active status of technologies by claiming that in the process of producing scientific facts we cannot take unambiguously extreme positions, which result from the modern critique of science.[28] Haraway emphasizes the instability of boundaries which define an object. According to her, this view seriously challenges the alleged modern division between nature and culture and calls into doubt the well-defined distinction between epistemology and ontology by claiming that the object of observation takes an active part in the process of knowledge production.[29] Barad postulates a better understanding of technologies as a space wherein nature and culture intra-act. This is where "matter and meaning meet," which means that practice should be included in the theory.[30]

In the case of the cinema, the discussion on the issue of visualization is exceptionally interesting on account of the fact that it is a medium that relies on visualization techniques: camera movements, the way the scenes are shot, and the narrative as such have long been regarded as spaces wherein ideology operates. Since every image is constructed in a very specific cultural context, and as such serves as a medium for the dominant meanings, then the overlap of science and popular culture becomes a focal point in the analysis of power relations. While contesting the value of visual evidence, contemporary science fiction tends to suggest that no forms of visual manifestation are either neutral or innocent. The SF genre has traditionally been seen as a suggestive and efficient means of spreading the idea of scientific and technological progress and an expression of the universal patriarchal construction of the Western culture as opposed to the effeminate nature. The identification of masculinity with reason and—analogically—of femininity with the body, combined with other binary oppositions, enabled the erasure of the role of the body as a subject of knowledge and, at the same time, resulted in the invasion of the issues connected to the body as an object of detached, objective, and detailed observation. Ian Hacking suggests that every look inside the

---

[26] Latour, *We Have Never Been Modern.*

[27] Grosz, *Volatile Bodies*; Haraway, *Modest_Witness@Second_Millennium.*

[28] Barad, "Meeting the Universe Halfway."

[29] Haraway, "Situated Knowledges."

[30] Barad, *Meeting the Universe Halfway.*

human body is also a transformation.[31] The process of seeing has an impact on the shape that the concept of the human body takes, and—consequently—influences its representations. The process encourages the creation of coherent transcendental models and schemes, grounded in the Western philosophy of negativity—as well as their oppositions—which triggers normativity originating from the Western idea of transparency. In order to finally break free from the dualistic logic of the original and the copy, the human and the inhuman, the same and the other, etc. we have to eventually start thinking not only *beyond* normativity but also *beyond* representation. Otherwise we will remain forever locked in the dualistic representational logic which offers no way out.

Works Cited

Balsamo, Ann. *Technologies of the Gendered Body*. Durham, NC: Duke University Press, 1996.

Barad, Karen. "Meeting the Universe Halfway: Realism and Social Constructivism without Contradiction." In *Feminim, Science, and the Philosophy of Science*, edited by Lynn Hankinson Nelson and Jack Nelson, 161–94. Dordrecht: Kluwer Press, 1996.

—. "Posthumanist Performativity: Toward an Understanding of How Matter Comes to Matter." *Signs: Journal of Women in Culture and Society* 28, no. 3 (2003): 801–31.

Baudrillard, Jean. *L'échange symbolique et le mort*. Paris: Gallimard, 1976.

—. "The Ecstasy of Communication." In *The Anti-Aestetic: Essays on Postmodern Culture*, edited by Hal Foster, 17–32. Port Townsend, WA: Bay Press, 1983.

Ben-Tov, Sharona. *The Artificial Paradise: Science Fiction and American Reality*. Ann Arbor: University of Michigan Press, 1995.

Berry, Philippa, and Andrew Wernik, eds. *Shadow of Spirit: Postmodernism and Religion*. London and New York: Routledge, 1992.

Bordo, Susan. *Unbearable Weight: Feminism, Western Culture, and the Body*. Berkeley: University of California Press, 1993.

Casper, Monica. "Reframing and Grounding Nonhuman Agency: What Makes a Fetus an Agent?"*American Behavioral Scientist* 37, no. 6 (1994): 839–56.

[31] Hacking, *Representing and Intervening.*

Crawford, Robert. "A Cultural Account of 'Health': Control, Release and the Social Body." In *Issues in the Political Economy of the Health Care*, edited by John McKinlay, 60–103. New York: Travistock, 1984.

Diamond, Irene, and Gloria F. Orenstein, eds. *Reweaving the World: The Emergence of Ecofeminism*. San Francisco: Sierra Club Books, 1990.

Dijck, José van. *Transparent Body: A Cultural Analysis of Medical Imaging*. Seattle and London: University of Washington Press, 2005.

Foucault, Michel. *The Birth of the Clinic: An Archeology of Medical Perception*. London: Travistock, 1973.

—. *Discipline and Punish: The Birth of the Prison*. London: Allen Lane, 1977.

Grosz, Elizabeth. "Bodies and Knowledges: Feminism and the Crisis of Reason." In *Feminist Epistemologies*, edited by Linda Alcoff and Elizabeth Potter, 187–216. New York and London: Routledge.

—. *Volatile Bodies: Toward a Corporeal Feminism*. Sydney: Allen and Unwin, 1994.

Hacking, Ian. *Representing and Intervening: Introductory Topics in the Philosophy of Natural Science*. Cambridge: Cambridge University Press, 1995.

Haraway, Donna J. "A Cyborg Manifesto: Science, Technology and Socialist Feminism in the Late Twentieth Century." In *Simians, Cyborgs, and Women: The Reinvention of Nature*, 149–81. London: Free Association Press, 1991.

—. *Modest_Witness@Second_Millennium.FemaleMan©_Meets_Onco Mouse™: Feminism and Technoscience*. New York: Routledge, 1997.

—. "The Promises of Monsters: A Regenerative Politics for Inappropriate/d Others." In *Cultural Studies*, edited by Lawrence Grossberg, Cary Nelson, and Paula Treichler, 295–338. London and New York: Routledge, 1992.

—. "Situated Knowledges: The Science Question in Feminism as Site of Discourse on the Privilege of Partial Perspective." *Feminist Studies* 14, no. 3 (1988): 575–99.

Harding, Sandra. *The Science Question in Feminism*. Ithaca, NY: Cornell University Press, 1986.

Keller, Evelyn Fox, and Christine Grontowski. "The Mind's Eye." In *Discovering Reality: Feminist Perspectives on Epistemology, Metaphysics, Methodology, and Philosophy of Science*, edited by Sandra Harding and Merrill B. Hintikka. Dordrecht: Reidel, 1983.

Latour, Bruno. *Science in Action*. Cambridge, MA: Harvard University Press, 1987.

—. *We Have Never Been Modern.* Cambridge, MA: Harvard University Press, 2002.

Marx, Leo. *The Machine in the Garden: Technology and the Pastoral Ideal in America*. London: Oxford University Press, 1964.

Rouse, Joseph. "Barad's Feminist Naturalism." *Hypatia* 19, no. 1 (2004): 142–61.

Springer, Claudia. "Psycho-cybernetics in Films of the 1990s." In *Alien Zone II: The Spaces of Science-Fiction Cinema*, edited by Annette Kuhn, 203–20. New York and London: Verso, 1999.

Turney, Jon. *Frankenstein's Footsteps: Science, Genetics and Popular Culture*. New Haven, CT: Yale University Press, 2000.

TRANSGENERIC TENDENCIES AND MULTIMEDIAL GENRES IN THE POETRY OF T. S. ELIOT AND THE FRENCH SYMBOLISTS

ALICJA PIECHUCKA

T. S. Eliot's *oeuvre* strikes both readers and critics as being suspended between the past and the present. On the one hand, his is a poetry permeated by a deep sense of the past and shaped by what is arguably the most interesting concept of tradition devised by any twentieth-century author or critic. It is common knowledge that Eliot's work thrives on inspiration which the poet found in distant epochs, ranging from antiquity to the seventeenth century in England and the nineteenth century in France. And yet this poetry steeped in the past is perceived as strikingly modern not only by some of Eliot's more progressive contemporaries, but also by today's readers and critics such as Richard Schusterman,[1] who refers to Eliot as a postmodernist.

One aspect of Eliot's poetry which seems particularly relevant to modern culture and the modern consciousness is his tendency to break down barriers between various literary and artistic genres. This tendency is part of Eliot's lifelong poetic quest for comprehensiveness and universality, which he attempts to achieve by combining seemingly disparate elements on the level of both form and content. In search of a comprehensive artistic vision of the world, Eliot marries innumerable literary genres and techniques, gravitating towards a fusion of various artistic media: not just literature, but also music and the visual arts. In doing so, he is largely inspired by the French symbolist poets. He therefore looks back on the tradition of nineteenth-century French poetry to achieve, paradoxically, what must be considered the summit of modernity. With his symbolist-inspired multifaceted and transgeneric poetry, Eliot approaches the postmodernist concept of multimedial genres and foreshadows the technological reality of our age at a time when communication technology as we know it could hardly have been conceived of. Eliot's desire to

[1] Leitch, *Norton Anthology of Theory*, 1092.

capture the heterogeneity and diversity of the world makes him a poet who is not only thoroughly modern, but also ahead of his times. In this context, the concept of multimedial genres put forward by Edward Balcerzan in an essay published in *Teksty drugie*[2] in 1999 seems relevant. Balcerzan associates what he refers to as multimedial genres with a combination of various media and resources, with all sorts of collages and literary kaleidoscopes, incorporating not only the literary and the artistic, but also, as Balcerzan importantly puts it, "life itself." The term could thus be applied to those literary and artistic forms which, due to their multifaceted, comprehensive nature, express the essence of our times and reflect, in literature, the contemporary means of communication such as television or the Internet. In my paper, I examine the way Eliot, inspired by the French symbolists, uses contrast, collage and juxtaposition, achieving, to use Balcerzan's term, "a metaphor for modernity."

The concept of collage-like diversity is central to both Eliot's work and French symbolist poetry, and manifests itself in juxtapositions of contradictory and dissonant elements, poetic combinations of the lofty and the low, high and popular culture, the comic and the tragic, the realistic and the fantastic. Such techniques inevitably entail the use of contrast. In this way, the poets in question convey the existential fears aroused by modern reality, with its unsolvable and irreconcilable contradictions. This tendency resurfaces even in the poems' titles. Relevant examples from Eliot's work immediately come to mind: "The Love Song of J. Alfred Prufrock," "Preludes," "Rhapsody on a Windy Night" and "Conversation Galante." In all four cases, the choice of the title is a blatant cultural reference, evoking associations with the domains of poetry, music and painting in the reader's mind. Thus, "The Love Song" brings to mind the long tradition of love poetry and, by extension, of lyric poetry in general. "Preludes" and "Rhapsody" connote the lofty, open-ended, *impromptu* nature of romantic musical compositions. Finally, the term *conversation galante* denotes a flirt-like conversation between a beautiful lady and an elegant gentleman in picturesque surroundings, sending the reader back to the French tradition of *poésie galante* and *peinture galante*, the latter best exemplified by the rococo artist Antoine Watteau. However, when Eliot reinterprets the genres, the content of each poem remains in contrast with the connotations of its title.

An examination of symbolist poems in terms of their titles helps identify the likely source of Eliot's inspiration. Rimbaud's "Nocturne vulgaire" combines the lofty and the sophisticated, traditionally associated

[2] Balcerzan, "W stronę genologii multimedialnej," 7–10.

with the term *nocturne*—a musical genre favoured by romantic composers as well as an indication of the poem's nocturnal setting—with the down-to-earth and the common. The result is ironic contrast between the expected ambience of a soft piece of piano music and the poem's ordinary setting, a contrast similar to that achieved by Eliot in "The Love Song of J. Alfred Prufrock." The symbolists thus show Eliot the possibilities the choice of title offers. The abundance of ambivalent titles in Corbière's *œuvre* proves the point. The first half of the longish title "À la mémoire de Zulma, vierge folle hors barrière et d'un louis" suggests an elegy for a woman, while the second, in which a sum of money is mentioned, makes it clear that it is a mock-elegy commemorating a prostitute.

Neither in Eliot's nor Corbière's poetry does the use of contrast remain limited to a play with literary genres and conventions. Analyses of Eliot's poems have demonstrated the poet's habit of ironically juxtaposing cultural allusions with evocations of the cheap, the tawdry, the down-to-earth and, in extreme cases, the indecent or the repulsive. When, in *The Waste Land*, musical references are mixed with images of moral and sexual debasement, Eliot seems to be echoing Corbière's "Paris":

> Musique!—C'est le paradis
> Des mahomets et des houris,
> Des dieux souteneurs qui se giflent![3]

The contrast between the brawls of Parisian pimps and the paradise which awaits devout Muslims is reinforced by that between the virginity of the houris and prostitution thriving in the French capital. The beautiful houris of Muslim paradise reappear as Eliot's Thames-daughters, the banks of the river being, in *The Waste Land*, a place from which "The nymphs are departed."[4] Eliot too brings together the past and the present, focusing on human relationships as the best exemplification of the contrast between the two: thus, Shakespeare's Juliet in "Portrait of a Lady," Ariadne, Theseus' loving guide, in "Sweeney Erect" or, in *The Waste Land*, Tristan and Isolde's undying love and Queen Elizabeth's passion for Leicester are opposed to their incomparably less heroic and glamorous modern counterparts.

In Eliot's poem "A Cooking Egg," the contrast between the poetic and the real is taken further as the poetic is juxtaposed with the mercantile:

---

[3] Corbière, *Les amours jaunes*, 23.
[4] Eliot, *Complete Poems and Plays*, 42.

I shall not want Capital in Heaven
For I shall meet Sir Alfred Mond.

We two shall lie together, lapt
In a five per cent. Exchequer Bond.[5]

In "Dans le Restaurant," Eliot's speaker mentions that he gave the waiter "dix sous"[6] to pay for the toilet. In *The Waste Land*, such an intrusion of the factual and the business-like is even more striking, because it affects the visual aspect of the poem. Thus, the words of the publican at the end of "A Game of Chess" are capitalized for greater effect: "HURRY UP PLEASE ITS TIME."[7] "The Fire Sermon" features "Mr. Eugenides, the Smyrna merchant," whose pockets are "full of currants / C.i.f. London: documents at sight,"[8] evoking the mercantile world which later reappears in "Death by Water," when the drowned Phlebas forgets "the profit and loss."[9] The roots of the tendency may again be traced back to the French symbolists. In "I Sonnet," Corbière sends a humorous telegram to his muse, asking for help: "Télégramme sacré—20 mots.—Vite à mon aide. . . | (Sonnet—c'est un sonnet—) ô Muse d'Archimède!"[10] The last line of the poem makes the contrast even more explicit due to the "Sonnet—Attention!" combination and the sarcastic, mock-romantic juxtaposition of two similarly-sounding words: "lyre" and "délire."[11] Corbière enhances the ironic dimension of his poems by inserting practical information into them, as in the title "À la mémoire de Zulma, vierge-folle hors barrière et d'un louis," which combines the elegy for a prostitute with the name of a currency. In "Vésuves et Cie," an ironic mention of money—"on m'a *fait* cent francs"[12]—is juxtaposed with the supposed splendor of Vesuvius, and the poem's very title ironically combines the volcano's name with the French abbreviation for company, the equivalent of the English *& Co*. The poem's Italian setting brings to mind Eliot's use of a similar device in "Lune de Miel," also an "Italian" poem, whose protagonists are likely to find in Milan "la Cène, et un restaurant pas cher."[13] Thus, with mastery which reveals Corbière's influence, Eliot combines *The Last Supper* by

[5] Ibid., 27.
[6] Ibid., 32.
[7] Ibid., 41.
[8] Ibid., 43.
[9] Ibid., 46.
[10] Corbière, *Les amours jaunes*, 39.
[11] Ibid., 40.
[12] Ibid., 118.
[13] Eliot, *Complete Poems and Plays*, 29.

Leonardo da Vinci and an inexpensive restaurant. In "Bambine," Corbière ironically emphasizes the contrast between a sea-wolf and the city slickers he takes on cruises during which he shows them "Pour vingt sous: *L'OCÉAN! L'OCÉAN!! L'OCÉAN!!!*"[14] the capitalization reinforcing the clash between the vastness of the ocean and the small amount of money charged for looking at it.

The same practice is traceable in Rimbaud: into his "Villes" ("L'acropole"), a poem in which the overwhelming impressiveness of the modern metropolis comes to stand for the confusing nature of modern civilization, the French poet introduces matter-of-fact, tourist-guide-like details. Rimbaud's fantastic, oneiric evocations of the cityscape are comically interspersed with facts and figures, with references to sizes—"Ce dôme est une armature d'acier artistique de quinze mille pieds de diamètre environ"—and prices—"Quelques divans de velours rouge: on sert des boissons polaires dont le prix varie de huit cents à mille roupies."[15] The effect is not only to create humorous contrast, but also to reinforce the poem's overall theme: a depiction of the modern state of mind. One of the key features of modernity, Rimbaud seems to suggest, is the mingling of the sublime and the down-to-earth, the technical and the mercantile invading the psychological and spiritual spheres of life. Such might also be the intention behind Corbière's and Eliot's parallel use of the device: the inclusion of detailed information such as prices successfully depoetizes verse, at the same time suggesting a concealed and bitter longing for the purely poetic as well as regret at life becoming materialistic and tawdry. The fact that such elements are often capitalized only makes them more visible and contributes to the layout of the "new" poem undergoing a transformation parallel to the one to which its content is subject.

If references to facts and figures, technical and business terms stand out against the rest of the text, then equally conspicuous, if not more so, are the foreign expressions which the author of *The Waste Land* habitually inserts into his poems. Besides directly quoting the symbolists and manifesting a marked preference for French expressions and words of Romance rather than Anglo-Saxon origin, Eliot refers to other non-English-language authors and incorporates Italian, German, Latin and Greek phrases into his work.

Interestingly, Eliot's penchant for polyglot collages also has precedents in French symbolism. In Baudelaire's "L'Horloge," we find an example of

[14] Corbière, *Les amours jaunes*, 181.
[15] Rimbaud, *Poésies*, 226.

poetic multilingualism as the speaker imagines that his clock reminds him about the passage of time in various languages: "*Remember! Souviens-toi*, prodigue! *Esto memor!* | (Mon gosier de métal parle toutes les langues.)"[16] Corbière is in the habit of inserting, in a collage-like manner, foreign interjections into his poems: "*English spoken?*"[17] the speaker of "Après la pluie" asks. More important than the role that a particular foreign borrowing plays in a particular poem are, however, the overall implications of poetic multilingualism. The narrator of Nerval's *Aurélia* recalls his attempt to "représenter, par mille figures accompagnées de récits, de vers et d'inscriptions en toutes les langues connues, une sorte d'histoire du monde mêlée de souvenirs d'études et de fragments de songes."[18] This short statement is an accurate summary of what Eliot will later try to do in his poetic *œuvre*: to put together multilingual stories, verse, inscriptions, bits of knowledge gained at school and dreams to create a universal history of the world. Thus, many decades before Eliot, the French poet sees the collage technique, with particular emphasis on foreign-language insertions, as a means of achieving poetic universality and comprehensiveness of vision. In a brilliant essay on Eliot, the Polish critic Wacław Borowy aptly comments on the Anglo-American poet's tendency to manifest his cosmopolitan inclinations by constructing what is, in Borowy's words, a poetic Tower of Babel.[19]

It may thus be said that Eliot follows in his symbolist predecessors' footsteps and constructs kaleidoscopic poems, whose comprehensiveness and dynamism result from his use of contrast and juxtaposition. The aim of the technique is to introduce jigsaw-puzzle-like diversity as well as to heighten and intensify the message of the poem. Frequently, the aim is also to provide the poem's structure. Upon a closer analysis of "The Love Song of J. Alfred Prufrock," one discovers that the poem revolves around string-like successions of nouns, which are frequently enumerated with almost no pause for breath. Thus, the "streets," "hotels" and "restaurants"[20] the speaker passes in the first stanza give way to "the window-panes," "the pools," the "chimneys," "the terrace" and "the house" which "the yellow fog" encounters in stanza three. As the poem unfolds, Prufrock tries to boost his self-assurance by listing the fashionable items of clothing he is wearing—"My morning coat, my collar mounting firmly to the chin, | My necktie rich and modest, but asserted by

[16] Baudelaire, *Œuvres Complètes*, 1:81.
[17] Corbière, *Les amours jaunes*, 44.
[18] Nerval, *Œuvres*, 775.
[19] Borowy, "Wędrówka nowego Parsifala," 12.
[20] Eliot, *Complete Poems and Plays*, 3.

a simple pin"[21]—and then goes on to resignedly recite all the things "I have known": "the evenings, mornings, afternoons," "the voices," "the eyes," "the arms."[22] It is the "tea and cakes and ices" that prevent the speaker of "The Love Song" from forcing "the moment to its crisis," just as "the cups, the marmalade, the tea" and "the porcelain" delay his ultimate gesture of squeezing "the universe into a ball."[23] As the poem draws to a close, Eliot's catalogue becomes increasingly packed and hurried, culminating in the speaker's realization of his own helplessness:

> Would it have been worth while,
> After the sunsets and the dooryards and the sprinkled streets,
> After the novels, after the teacups, after the skirts that trail along the
>   floor –
> And this, and so much more? –
> It is impossible to say just what I mean![24]

Prufrock's dramatic list ends in a nervous, impatient, et caetera-like dismissal. One might wonder whether the poet has put a question mark after the phrase "so much more" merely because it results from the overall grammatical structure of the sentence or has constructed the sentence so that it is possible for him to question the apparent diversity and abundance of life, suggesting that "so much more" amounts in fact to very little. In resolving the matter, a comparison with French symbolist poetry might again prove helpful.

It is from Corbière more than any other French symbolist that Eliot has learned the art of poetic enumerations, of constructing elaborate poetic catalogues, which frequently constitute the basic structure of the poem, its backbone. A case in point is Corbière's "Épitaphe," which is in fact a two-page catalogue, a long poetic enumeration of what the poem's protagonist is and is not. So is "Litanie du sommeil," where on as many as six pages Corbière mostly lists various periphrases which refer to sleep, as well as the phenomena to which he compares sleep. It is obvious that the function of the catalogue-like enumerations in Corbière's poems is highly metaphorical: not only should the various elements not be taken literally, but together they stand for much more than just their sum. Basing the structure of the poem on such a string-like succession of elements enables Corbière to convey what is not just the essence of these particular poems,

[21] Ibid., 4.
[22] Ibid., 5.
[23] Ibid., 6.
[24] Ibid., 6.

but of his verse in its totality and perhaps modern verse in general. It is the overwhelming sense of confusion, rootlessness and ambiguity of the vulnerable self exposed to a dense, heterogeneous, incomprehensible and hostile reality, turning the individual into a tangle of contradictions. No other poet understands Corbière's message and makes use of it to such an extent as Eliot. To a greater or lesser degree, the device is used throughout Eliot's pre-conversion poetry, culminating in the collage structure of *The Waste Land*.

More important than its meaning is, however, the way in which enumeration replaces, through its elliptic, evocative character, lengthy descriptions and verbose complexities. By putting together various objects or, as here, people or groups of people, Corbière achieves the same effect as Eliot does with the objective correlative. The latter, "The only way of expressing emotion in the form of art," is "a set of objects, a situation, a chain of events which shall be the formula of that particular emotion; such that when the external facts, which must terminate in sensory experience, are given, the emotion is immediately evoked."[25] This is precisely what Corbière's enumerations are: strings of objects or people following one another in a chain-like succession, as if the poet wanted to save time and space by enumerating various elements, leaving it to the reader to make the connections and work out the emotions for himself. The French poet thus seems to share Eliot's belief that the "new" poetry should be difficult, and that the poet's task is to force the reader to actively participate in the poetic process by deciphering the hermetic structure of verse on his own. In an Eliotesque vein too, Corbière's consistent use of poetic catalogues is prompted by his fear of over-describing, which is perhaps the key factor underlying Eliot's theory of the objective correlative. In everyday speech, the use of ellipsis is aimed at saving time and constitutes a mental shortcut of sorts; in poetry, it adds dynamism and increases the tempo. Most importantly, however, it results in stripped-down, verbally economical and laconic verse, but also, paradoxically, renders its reception more complicated and challenging by requiring the reader to decode the meanings inherent in indirect, opaque poetic form. Thus, the poetry of both Eliot and the symbolists draws heavily on seemingly arbitrary amalgams of images and leaves it to the reader to make sense of the apparent disconnectedness.

Importantly, the symbolist-Eliotesque associative practice, relying on techniques such as juxtaposition and enumeration, turns out to be a means of evoking the modern comprehensiveness. On the other hand, Corbière's

[25] Eliot, *Sacred Wood*, 100.

use of catalogues makes it possible, so to speak, to throw everything into one basket, and thus achieve his favorite effect of reducing disparate phenomena to the lowest common denominator in order to demonstrate the general degradation and devaluation of modern life. By amassing within one stanza, or even one line, various elements, he implies their shared unimportance. In the light of Corbière's consistent use of the technique, the seemingly casual way in which Eliot lists various objects and series of actions in his poems acquires a new dimension. In this respect, some fragments from "Portrait of a Lady" are perhaps even more striking and relevant than the passages from "The Love Song" quoted earlier:

> Let us take the air, in a tobacco trance,
> Admire the monuments,
> Discuss the late events,
> Correct our watches by the public clocks.
> Then sit for half an hour and drink our bocks.
>
> You will see me any morning in the park
> Reading the comics and the sporting page.
> Particularly I remark
> An English countess goes upon the stage.
> A Greek was murdered at a Polish dance,
> Another bank defaulter has confessed.[26]

Following Corbière's example, Eliot manages to convey the nature of modernity, focusing on its two key aspects: the confusing multifariousness and the swift pace, which are mirrored, respectively, by the incongruousness and telegraphic shortening inherent in catalogue- and collage-like techniques. Like his French predecessor too, the Anglo-American poet is painfully aware that the former aspect fails to conceal the superficiality and tawdriness of contemporary reality, while the latter serves to unsuccessfully disguise the overwhelming sense of emptiness the modern Everyman experiences.

Much has been said about the heterogeneous, collage-like structure of *The Waste Land*, with its whirlwind of people and places, allusions and quotations, voices, styles and moods, which make Eliot's magnum opus an unprecedented poetic mélange of the contrasting and the contradictory, a unique mixture of reality and fantasy, history and mythology, psychology and religion. The result is a mysterious, uncontextualized, seemingly irrational and, at first glance, nearly absurd amalgam, which obliges the puzzled reader to work out the semantic and logical connections by

[26] Eliot, *Complete Poems and Plays*, 9–10.

himself. If some technical aspects of Eliot's earlier poems point to his indebtedness to the author of *Les Amours jaunes*, the poetic shape of *The Waste Land* may undoubtedly be traced back to Rimbaud, who takes the use of the catalogue technique further than Corbière.

Thus, in part three of "Vies," Rimbaud's speaker enumerates various places where he has gained knowledge and accomplished various things, juxtaposing disparate experiences and locations. Another poem, "Promontoire," perfectly exemplifies Rimbaud's penchant for enumerative poetic constructs. The three opening lines refer to a voyage, showing a ship as it reaches a promontory. The rest of the poem details what is situated there in one longish sentence. Semicolons and dashes separate the clauses, making the poem even more list-like and reducing each element of Rimbaud's poetic vision to the level of an item on that list. The disconnectedness thereby achieved, which, in a do-it-yourself vein, leaves it to the reader to establish the links, reflects the disparate nature of the elements Rimbaud evokes: "l'Épire et le Péloponnèse," "la grande île du Japon," "l'Arabie," "de grands canaux de Carthage," "des Embankments d'une Venise louche," "de molles éruptions d'Etnas," "de peupliers d'Allemagne," "des têtes d'Arbres du Japon," "les façades circulaires des « Royal » ou des « Grand » de Scarborough ou de Brooklyn."[27] What all the elements have in common is, of course, that they are geographical locations. Apart from that, however, everything seems to separate them. The distance between them is not only spatial, but also temporal: the Peloponnese and Carthage, the latter being, incidentally, one of Eliot's favorite poetic locations, take the reader back to antiquity, while the hotels of New York or English seaside resorts, again bringing to mind Eliot's "Cannon Street Hotel" and "the Metropole"[28] in *The Waste Land*, are deeply rooted in the present. The very grammatical implications of the catalogue structure contribute to the impression of timelessness. The flow of nominal phrases, interrupted only by semicolons and dashes, makes it possible to reduce the use of verbs: as a result, tenses are largely eliminated, and temporal references seem to have been obliterated. The past and present overlap, and so do East and West. While Europe merges with Asia, nature fuses with culture, Italian volcanoes, German poplars and Japanese trees emerging amid the architecture of the world's most famous cities. The result is an impression of fullness, vastness and vertiginous multitude.

---

[27] Rimbaud, *Poésies*, 244.

[28] Eliot, *Complete Poems and Plays*, 43.

Like Eliot in *The Waste Land*, though on a smaller scale, Rimbaud achieves in "Promontoire" a condensation of not only the globe, but of history, civilisation, human consciousness and the very essence of life. By putting together various cities, countries and continents, various architectural constructions and elements of familiar and exotic flora, he achieves an all-embracing vision, covering all aspects of the modern world, whose map is traced in a universalized vision. The result is Rimbaud's and Eliot's peculiar poetic geography, at once magical and realistic, time-bound and non-temporal. What both poets attempt to do is to adjust the poetic enterprise to the challenges of the *Zeitgeist*, to write poems whose form and scope correspond to the multifariousness of the era. The enterprise is as maddeningly impossible as the times themselves: when Rimbaud speaks of "des plus colossales constructions de l'Italie, de l'Amérique et de l'Asie,"[29] he could just as well be referring to the modern poetic challenge, as colossal as the monumental works of modern architecture. The catalogue-like construction is a response to the tempo of modern existence: hence the choice of a hasty, almost telegraphic form, a shortcut to conveying the poetic message. The burden of making sense of it is shifted onto the reader, rendering modern poetry another source of confusion, in addition to the one provided by modern life.

Eliot's famous essay "The Metaphysical Poets" praises Donne and Cowley for their ingenious use of the conceit, in particular their capacity for "a development by rapid association of thought which requires considerable agility on the part of the reader."[30] The success of Donne's poems is, according to Eliot, frequently "secured by brief words and sudden contrasts,"[31] the latter being specified as "the sudden contrast of associations."[32] Juxtaposition, contrast and diversity are the hallmarks of true poetic achievement: "a degree of heterogeneity of material compelled into unity by the operation of the poet's mind is omnipresent in poetry,"[33] Eliot asserts, supporting his view with a quotation from Baudelaire's "Le Voyage" and later with references to Laforgue and Corbière. The metaphysical poets and the French symbolists have managed to avoid the disastrous "dissociation of sensibility"[34] which Eliot deplores, because the associative faculties they possess enable them to create an emotional-intellectual poetic complex, in which "a direct sensuous apprehension of

---

[29] Rimbaud, *Poésies*, 244.
[30] Eliot, *Selected Essays*, 242.
[31] Ibid., 242.
[32] Ibid., 243.
[33] Ibid., 243.
[34] Ibid., 247.

thought" and "a recreation of thought into feeling"[35] are possible. Through "a contrast of ideas"[36] and "the richness of association,"[37] "the idea and the simile become one."[38] The inclusion in verse of a variety of seemingly unrelated images translates itself into the wide spectrum of thoughts and feelings such poetry manages to convey. Unlike "the ordinary man's experience," that of the true poet is not "chaotic, irregular, fragmentary": love, Spinoza's writings, distracting noises and kitchen smells are brought together to form "new wholes." The mind of the poet possessed of an associative sensibility "could devour any kind of experience," whether it is "simple, artificial, difficult, or fantastic."[39] The comprehensive nature of the poem allows for a better reflection of the complex, multifaceted nature of the world and the modern man's psyche.

In this way, the supreme aim of French symbolism is attained: a universality of poetic expression, in which unity and multifariousness are reconciled. Moreover, the hermetic, esoteric quality of verse is amplified, as the composite, assemblage-like poetry for which Eliot and the symbolists argue is properly challenging and demanding. It is up to the reader to work out the "connexions which are not implicit,"[40] to use Eliot's words. In this versatile, pluralistic and universal poetry, the vital is left unsaid: just as the poet ought to be possessed of an associative mind, the reader needs to be equipped with the ability to read between the lines. Most strikingly, however, the application of collage-like poetic techniques, in addition to realising Eliot's ideal of an associative sensibility, is also related to another key notion put forward by the symbolist school: the correspondence of the arts. Collage is, it must be remembered, a technique used in the visual arts. It consists in pasting various materials or objects onto a surface: it is thus a painterly term adopted by literary theorists. Appropriately enough, the aesthetic notion of *la correspondance des arts* postulates the existence of analogies, relationships and parallels between various, seemingly dissimilar branches of art. As early as the first century B.C., Horace coins the famous formula "Ut pictura poesis," "As in painting, so in poetry." The idea is fully developed by the French symbolist poets, who extend its scope and meaning. "Correspondances," Baudelaire's sonnet and poetic manifesto, shows that various arts not only share the same sources of inspiration, but the impressions created by one

[35] Ibid., 246.
[36] Ibid., 243.
[37] Ibid., 244–45.
[38] Ibid., 243.
[39] Ibid., 247.
[40] Ibid., 242.

art are also translatable into those provided by another. In brief, the central belief behind the *correspondance des arts* concept is that there exists a network of correspondences between literature, music and the visual arts, just as there exist connections between various kinds of sensory impressions. When the symbolist poets experiment with collage, contrast and juxtaposition, they anticipate *The Waste Land* as much as the collages by Georges Braque and Pablo Picasso, thereby initiating a quest for completeness which presents the modern artist with the near-impossible task of achieving plenitude and unity while embracing the heterogeneous reality of the modern world.

Works Cited

Balcerzan Edward. „W stronę genologii multimedialnej." *Teksty drugie*, no. 6 (1999): 7–24.

Baudelaire, Charles. *Œuvres complètes.* Edited by Claude Pichois. 2 vols. Paris: Editions Gallimard, coll. Bibliotèque de la Pléïade, 1972.

Borowy, Wacław. "Wędrówka nowego Parsifala." *Przegląd Współczesny* 15 (1936): 10–16.

Corbière, Tristan. *Les amours jaunes.* Paris: Gallimard, 1973.

Eliot T. S. *The Complete Poems and Plays*. New York: Harcourt, Brace and World, 1962.

—. *Sacred Wood*. London: Methuen, 1972.

—. *Selected Essays*. New York: Harcourt, Brace and World, 1960.

Leitch, Vincent, ed. *The Norton Anthology of Theory and Criticism*. New York: W. W. Norton, 2001.

Nerval, Gérard de. *Œuvres*. Paris: Garnier Frères, 1958.

Rimbaud, Arthur. *Poésies*. Kraków: Wydawnictwo Literackie, 1993.

NOSTALGIC AVANT-GARDES: WRITING POETRY IN THE AGE OF E-MAIL

JULIA FIEDORCZUK

Is there a place for poetry in a computerized world? In what way do the new information technologies influence how poetry gets written, distributed and read? What differentiates experimental poetry from electronic spam? How does one distinguish a "genuine" avant-garde work of art from a commodity that mimics avant-garde gestures? Is Marjorie Perloff right in claiming that L=A=N=G=U=A=G=E poetry is our contemporary poetic avant-garde? Before one attempts to answer these questions, it is necessary to clarify the basic terms of the argument.

The term avant-garde and its relation to modernism is notoriously difficult to theorize. There is no space here for more than a cursory treatment of this problem. Most theoreticians point to a split within modernism into its "high" (post-symbolist) aspect and its avant-garde agenda. In this scheme, "High Modernism" represents the redemptive and solidifying ambitions of the moment, whereas avant-garde epitomizes its destabilizing, revolutionary, or anarchic tendencies.

This, however is much too imprecise a map. The two strains are not easily told apart since both "High Modernism" and the avant-garde, or the "Other Tradition," are modes of creation through rupture and crisis. In *Five Faces of Modernity*, Mattei Calinescu argues that the split consciousness, and therefore the conflict between revolutionary aspirations on the one hand and totalitarian impulses on the other, is characteristic of modernity as such and can be observed in all of its "faces" (modernism, the avant-garde, decadence, kitsch, postmodernism). In Baudelaire, for instance, this divided consciousness assumes the form of a clash between a revolutionary rejection of normative past (his characterization of modernity in "The Painter of Modern Life" as the realm of the transient and the fugitive) and a kind of nostalgia, a dislike of the vulgar present.

This conflict must become conscious, if a truly modern aesthetic is to be achieved.[1]

Writers attempting to define the avant-garde focused on the following characteristics: avant-garde art's opposition to kitsch (Greenberg) or mass culture (Adorno, Horkheimer), its rejection of both traditional and current cultural trends (Poggioli), the alienation of the artist and his or her rejection of bourgeois values. A mapping of these problems has been presented by Peter Bürger in his *Theory of the Avant-Garde* (1984). Bürger's work very effectively distinguishes the avant-garde from other aspects of modernism, and exposes the profoundly paradoxical nature of avant-garde projects. His theory will be used to cast light on several theoretical problems connected with the activities of L=A=N=G=U=A=G=E poets, represented, for the sake of this argument, by Charles Bernstein and Ron Silliman.

Bürger focuses his argument on the differences between Aestheticism and the avant-garde. In Jochen Schulte-Sasse's formulation, "Aestheticism's intensification of artistic autonomy and its effect on the foundation of a special realm called aesthetic experience permitted the avant-garde to clearly recognize the social inconsequentiality of autonomous art and, as the logical consequence of this recognition, to attempt to lead art back into social praxis."[2] This dimension of social praxis is what effectively distinguishes avant-garde from aesthetic modernism: "Modernism may be understandable as an attack on traditional writing techniques, but the avant-garde can only be understood as an attack meant to alter the institutionalized commerce with art."[3] Avant-garde negates the autonomy of art characteristic of the bourgeois society and this negation occurs through the questioning of the very status of art:

> When the avant-gardistes demand that art become practical once again, they do not mean that the contents of works of art should be socially significant. The demand is not raised at the level of the contents of individual works. Rather, it directs itself to the way art functions in society, a process that does as much to determine the effect that works have as does a particular content.[4]

---

[1] Calinescu, *Five Faces of Modernity*, 58.
[2] Schulze-Sasse, foreword to *Theory of the Avant-Garde*, by Bürger, XIV.
[3] Ibid., XV.
[4] Bürger, *Theory of the Avant-Garde*, 49.

But the social praxis into which the production of an artwork must be reintegrated is not what Bürger calls "the means-ends rationality of the bourgeois everyday."[5] In turning away from this praxis of life avant-garde art agrees with aestheticism. However, this is where the two tendencies part company, the aim of the avant-garde being "to organize a new life praxis from a basis in art."[6]

This is a completely paradoxical, or indeed an impossible endeavor, because a critical perspective on a given social praxis is only possible when art is granted a certain amount of autonomy, and that is precisely what the avant-garde negates. Without distancing itself from the bourgeois praxis of life, avant-garde art may easily become conflated with commodity aesthetics. We are then dealing with, in Bürger's terms, a "false sublation of art as institution." Bürger concludes: "Given the experience of the false sublation of autonomy, one will need to ask whether a sublation of the autonomy status can be desirable at all, whether the distance between art and the praxis of life is not requisite for that free space within which alternatives to what exists have become conceivable."[7]

Paradoxically then, art can (theoretically) be socially significant only if it distances itself from the social praxis of humans, which, in turn, prevents it from being integrated with that praxis. I believe that this crucial inconsistency constitutes the very core of the experimental project of L=A=N=G=U=A=G=E poetry, considered by many authors, notably by Marjorie Perloff, as the contemporary literary avant-garde. I claim that, if works by Charles Bernstein and other poets associated with the L=A=N=G=U=A=G=E school have any social significance whatsoever, it is only insofar as they depart from at least some of their avant-garde assumptions. Perloff's sympathetic analysis of contemporary experimental poetry—for instance in her book *Radical Artifice: Writing Poetry in the Age of Media*—seems to pass over certain crucial theoretical problems, notably what Paul Mann has termed as "theory death" of avant-garde (its dissolution in the mass of theoretical discourse it generates) and the reactionary undercurrent of nostalgia present in many experimental projects.

Perloff claims that the difficulty and non-transparency of some postmodern American poetry is a reaction against the formulaic character of much of our everyday communication. The author's example of this super-conventionality is a stationary line called EXECARDS, designed (as we learn from an advertising leaflet reproduced in Perloff's book) to

---

[5] Ibid.

[6] Ibid.

[7] Ibid., 54.

ensure successful communication in a number of business and personal situations. EXECARDS are divided according to categories such as: thanking, acknowledging, reminding, rewarding, prospecting, following up, welcoming and inviting. All one needs to do is buy a postcard and get the message across, quickly and effortlessly, no noise involved.

Such reduction of human communication to sellable formulas is not unrelated, in Perloff's view, to the fact that we live in a technologically advanced world. Computerization, especially, has a far-reaching impact on the forms of human exchange. Interaction with a computer involves the negation of temporality and variability. Desired feedback is obtained by pressing the right key. In Perloff's words, "One either calls up a file or one doesn't, without going through a series of linear or temporal steps."[8] Thus, "communication" with a computer involves a binary choice, and all shades of meaning must be eliminated. The shady space in between, the layers of language not used, to allude to Wittgenstein, in the language game of exchanging information, are brought to the foreground in the experimental practices of Perloff's favorite poets. That, in this critic's opinion, makes these poets' work subversive and endows it with political significance. In Perloff's view, the radicalism of contemporary avant-garde is a consequence of its formal non-conformism.

However, one might observe that the political potential of these works can only be activated if one reads them against their avant-garde ambitions, disregarding some of the self-conscious claims these works make about their own status. This necessary misreading involves such interpretive strategies as reconstructing the missing coherence of a text, establishing connections between disparate fragments, or contextualizing its experimental devices. "A Defense of Poetry" by Charles Bernstein will serve as the first example. In order to appreciate the subversive flair of this poem one has to submit it to an interpretation which involves placing it in a theoretical framework that will help to explain away its formal oddities. Here is the beginning of the poem:

> My problem with deploying a term liek
> nonelen
> in these cases is actually similar to
> your
> critique of the term ideopigical
> unamlsing as a too-broad unanuajce
> interprestive proacdeure.
> You say too musch lie a steamroller when

[8] Perloff, *Radical Artifice*, 188.

> We need dental (I'd say jeweller's)
> tools.
> (I thin youy misinterpret the natuer of
> some of the poltical claims go; not
> themaic
> interpretatiomn of evey
> evey detail in every peim
> but an oeitnwtation towatd a kind of
> texutal practice
> that you prefer to call 'nknsense' but
> for political purpses I prepfer to call
> ideological!
> ,say Hupty Dumpty)[9]

The poem's orthographic irregularities are an extension of its content. The text performs a critique of the meaningless discourses of dominant politics (Bush's speeches are quoted as an example) and the mass-media (the media-ization of language). The text sets out to deconstruct the binarism of sense and non-sense and invent a new way of speaking. It attracts the reader's attention to the materiality of the signifier, thus spelling out a warning against identifying language with reality. It sets itself apart from the ideal of transparent meaning, perceived by some of the L=A=N=G=U=A=G=E authors as an extension of commodity fetishism characteristic of consumer capitalism. But all this subversive activity can only be appreciated if one ignores the idiosyncratic orthography of the poem and identifies pieces of perfectly coherent argument; it is not very difficult to do, since the majority of misspelled words are easily recognizable. The following fragment will illustrate my point:

> . . . Indeed you say that
> nonsense shed leds on its 'antithesis'
> sense making: but teally the antithisi
> of these poems you call nonselnse is not
> sense-making itself but perhps, in some
> cases, the simulation of sense-making:
> decitfullness, manifpulation, the
> media-ization of language, etc.[10]

When one reconstructs the correct spelling and punctuation and eliminates line-breaks, the fragment reads as follows: "Indeed you say that nonsense sheds light on its 'antithesis,' sense-making: but really, the antithesis of

---

[9] Bernstein, *My Way*, 1.
[10] Ibid., 2.

these poems you call nonsense is not sense making itself but perhaps, in some cases, the simulation of sense-making: deceitfulness, manipulation, the media-ization of language, etc." According to the argument presented here, public discourse is "a simulation of sense-making" because of its hyperconventional correctness. The poem contests this linguistic mystification. But is experimental orthography a noteworthy alternative to the totalitarian speech of the representatives of global capitalism? In Bob Perelman's words, Bernstein's solution is "the utopia of the liberated signifier, the sagnifire, the snugifirm, the nugiform."[11] The solution is utopian because signifiers can never be totally divorced from the things they are meant to signify, and even if that were possible, the usefulness of such a split-up for oppositional politics seems highly questionable. Besides, as Perelman notes further on, the speech of the politicians, criticized by Bernstein, hardly ever possesses the clarity the poet attributes to it. Perelman writes: "Recent books . . . quoting Bush verbatim reveal how random and deteriorated his actual speech can be, but clearly these errata promise scant political liberation. And if the target that Bernstein is attacking is a mirage of convention, the utopian politics of liberated textuality are also cloudy."[12]

There is nothing inherently oppositional about orthographic incorrectness. One ought to note that the errors incorporated into Bernstein's poem are, in most cases, simple typos. If there were something essentially subversive in such typing mistakes then all of us—in the age of e-mail and cell phones—could consider ourselves revolutionists. That is clearly not the case, and careless spelling could just as well be perceived as a token of compliance with the demands of capitalist economy which values speed of communication more than it does the quality of communicated contents. Therefore, "A Defense of Poetry" only preserves its radical potential when divorced from the everyday life-praxis of technologically-informed society and read in a special way, in an institutionalized context (academy). In Peter Bürger's terms, it must be granted a certain amount of autonomy, which of course goes against its avant-garde status. But only then can it be read as a poem—and not spam. Contemporary avant-garde thus seems to be caught up in what Paul Mann termed as its "theory death."[13] The history of L=A=N=G=U=A=G=E writing yields substantial evidence to this claim. As Barrett Watten noted in his essay "The Secret History of the Equal

---

[11] Perelman, "Write the Power," 317.

[12] Ibid., 318.

[13] For the discussion of Mann's analysis of the "theory death" of the avant-garde, see Watten, "Secret History of the Equal Sign," 581–627.

Sign: L=A=N=G=U=A=G=E between Discourse and Text," the journal L=A=N=G=U=A=G=E, edited by Charles Bernstein and Bruce Andrews, was mostly devoted to the statements of poetics, examples of language-centered writing being of secondary importance.[14]

If L=A=N=G=U=A=G=E fails, as I think it does, to challenge the autonomy of the literary work, it is perhaps more successful at surmounting the autonomy of the authorial subject. Watten cites *Legend* (1980), written collectively by Bruce Andrews, Charles Bernstein, Ray DiPalma, Steve McCaffery, and Ron Siliman, as an example of dialogic, collective writing. The text contains sections written by individual authors, as well as fragments co-written by two or more persons. Watten notes that the single-author sections "provide an opening orientation and a thematic continuity that the work progressively moves away from, toward a final horizon of collective authorship."[15]

Watten theorizes this dismantling of individual authorship in terms borrowed from Kristeva's *Revolution in Poetic Language* and Foucault's *Archeology of Knowledge*. The shattering of "the bourgeois ego" performed in *Legend* is supposed to open up the space for revolutionary subjectivity. This radical subjectivity, or rather inter-subjectivity, opposes the stratification of language in capitalism by resorting to collectivity. This gathering of diverse subjects united in their defiance of hegemonic forms of discourse is held together by nostalgia. The utopian dream it invokes is that of a community: a democratic forum where voices mix and blend, forming an unbound intertextuality. Watten, who does not speak of nostalgia in connection with L=A=N=G=U=A=G=E experiments, notes that the community which the writing of *Legend* initiates is markedly phallogocentric. I quote from a section authored by Ron Silliman:

> 11. It is tribal.
> 12. It is male.
> 13. It is behavior.

---

[14] Watten, "Secret History of the Equal Sign," 585–87. Watten is of course aware of the intention, on the part of some L=A=N=G=U=A=G=E authors, to eliminate the generic specificity of their writing. Hence numerous attempts at combining poetry and exposition of poetics within a single text. However, Watten believes in the continuing relevance of genre. He writes: "if language writing indeed successfully produces texts that do not have generic specificity, the implications for the conception of the literary are large. The literariness that results would be, in my view, transcendental; to maintain the historical specificity of literary production, it is necessary to have a notion of genre seen as produced in specific contexts" (594).

[15] Ibid., 597.

> 14. It does not conceal.
> 15. It shares the labor but does not divide it.
> 16. It could do anything.
> 17. It is a poem.[16]

It would be a mistake to overlook Silliman's irony, or to identify the "it" that begins each of the quoted sentences with "a poem," as the last line encourages us to do. Nevertheless, what we get in this fragment is a complete set of avant-garde paraphernalia: whatever "it" is (and one may assume that "it" is a point of this writing), it is opposed to closure, alienation, and injustice. It is "tribal," and, like Lewis and Pound's vortex, it is also "male." This last denotation could simply be a reference to the fact that *Legend* was authored by five men but it does not make the fragment any less nostalgic. Silliman imagines a community of men united in "tribal" activities. The labor is shared but not divided, and this togetherness makes anything possible.

The section authored by Bernstein is not phallogocentric but its undercurrent of nostalgia is even more evident. He writes: "in the beginning was one & we can remember it by looking inside ourselves, into (back to) our self-sameness. & the language of us is the body of this one," and, further on, "the words are the scent of our usness"[17] (127). This is strongly evocative of Laura (Riding) Jackson's *The Telling*, the work which that modernist poet considered as her "personal evangel" and which was a culmination of her life-long search for a truthful form of expression. In the Outline preceding the body of *The Telling*, (Riding) Jackson voiced her dissatisfaction with the compartmentalization of human thought and language resulting from the egoistic practices of "self-claiming selves." (Riding) Jackson grounded her belief in the redemptive potential of language in a conviction that a different kind of selfhood is possible:

> *There is an alternative self, a human-souled being, a self conscious of ourselves who bear in manifold individualness, each singly, the burden of the single sense of the manifold totality. This self is implicated in the totality as a speaking self of it, owing it words that will put the seal of the Whole upon it* [italics in the original].[18]

Both (Riding) Jackson and Bernstein use neologisms such as "usness" (Bernstein) or "individualness" (Riding) and both refer to some mythical

---

[16] Andrews et al., *Legend*, 14.
[17] Ibid., 127.
[18] Jackson, *Telling*, 3.

moment in the history of the human subject, or indeed of humanity as such, when language was not an alienating factor and human beings experienced themselves as fundamentally connected with the rest of the universe. For both writers, language contains the traces of that prelapsarian (pre-Oedipal, pre-capitalist) harmony, though they devise very different strategies for activating this redemptive potential of language. Both projects turn to the past in order to find there the promise of a better future.

Legend is but one example of collective authored texts that came out of the L=A=N=G=U=A=G=E school and certainly the conclusion presented here need not apply to all of them. I am tempted to suggest, however, that a kind of nostalgia, or what Andrzej Sosnowski denoted as a "retroactive hope,"[19] is an important aspect of many collective experimental undertakings, since the liberating new communities they posit are almost always on some level a return to a more "natural state," even if the means of this return is provided by advanced technology.

Poetics listserv, founded by Charles Bernstein in 1994, is an experimental project which makes use of the new forms of communication offered by the Internet. It started as a private e-mail discussion group, then grew into a large, moderated forum, devoted to an ongoing discussion of experimental poetics. One of the first discussions on "Poetics" concerned the subject of "poetry communities," especially in the context of the newly-emergent notion of "virtual communities" made possible by cyberspace. In his posting (later published as "Community and the Individual Talent"), Bernstein is very critical of the notion of community and very much aware of the fact that every community will be a site of exclusion before it becomes the site of bonding. His hopes, with cyberspace, are for what he calls "virtual uncommunities." He writes:

> I want to imagine reading and writing, performing and listening, as sites of conversation as much as collectivity. I want to imagine a constellation of readers who write, to and for one another, with the links always open at the end, spiraling outward—centrifugally—not closing in.[20]

This certainly sounds less "tribal" than Bernstein's contribution to *Legend*, but not less utopian. In fact, links are not always open, and perhaps the Internet has also severed some of them, even as it opened up new possibilities of connection (for instance because some readers of poetry are computer-illiterate or have no access to the Internet). Furthermore,

---

[19] Sosnowski, "Posłowie," 178.
[20] Bernstein, "Community and the Individual Talent," 179.

Bernstein's imagined "uncommunity" is in fact a community par excellence, free of the imperfections of olden-day, un-virtual communities: a community which does not rely on exclusion, which is open-ended and always growing.

To conclude, it must be noted that Bernstein is perfectly aware of the problems discussed in the present article. If his writing does not always successfully evade the aporias of the avant-garde, its strength is that it can look upon its own limitations and laugh. Perhaps humor saves his work from "theory death" and from becoming nostalgic, but at the same time it marks its distance from serious, avant-garde ambitions. Bernstein's experimental work is thus successful in so far as it questions the very rationale of its experimentation.

Works Cited

Andrews, Bruce, Charles Bernstein, Ray DiPalma, Steve McCaffery, and Ron Silliman, *Legend*. New York: Segue Foundation, 1980. http://english.utah.edu/eclipse/projects/LEGEND/legend.html (accessed March 5, 2009).

Bernstein, Charles. "Community and the Individual Talent." *Diacritics* 26, no. 3/4 (1996): 176–95.

—. *My Way: Poems and Speeches*. Chicago, London: University of Chicago Press, 1999.

Bürger, Peter. *Theory of the Avant-Garde*. Translated by Michael Shaw. Foreword by Jochen Schulze-Sasse. Vol. 4, *Theory and History of Literature.* Minneapolis: University of Minnesota Press, 1984.

Calinescu, Mattei. *Five Faces of Modernity*. Durham: Duke University Press, 1987.

Jackson, Laura (Riding). *The Telling.* London: Athlone Press, 1972.

Perelman, Bob. "Write the Power." *American Literary History* 6, no. 2 (1994): 306–24.

Perloff, Marjorie. *Radical Artifice: Writing Poetry in the Age of Media.* Chicago and London: University of Chicago Press, 1991.

Sosnowski, Andrzej. "Posłowie." Afterword to *Pieśni*, by Ezra Pound, 175–87. Warszawa: PIW, 1996.

Watten, Barret. "The Secret History of the Equal Sign: L=A=N=G=U=A=G=E between Discourse and Text." *Poetics Today* 20, no. 4 (1999): 581–627.

ELECTRONIC FLOATING OPERA AND CHALLENGES OF FICTIONAL NAVIGATION IN JOHN BARTH'S NOVEL *COMING SOON!!! A NARRATIVE*

MARTA KOVAL

Concluding the epilogue to the 2002 edition of her almost classic *The Politics of Postmodernism,* L. Hutcheon remarked:

> Electronic technology and globalization, respectively, have transformed how we experience the language we use and the social world in which we live. For many, these changes are simply other manifestations of postmodernity. But what if we consider these as the first signs of what will come *after* [italics in the original] the postmodern? The intertextual, interactive aesthetic suggested by hypertextuality is related to the postmodern, to be sure, but is it the same thing? What if postmodern parody were merely the preparatory step to a "Net" aesthetic, utopianly defined as a "non-linear, multivocal, open, nonhierarchical aesthetic involving active encounters"? And what about the visual as well as verbal dimensions of electronic creation? . . . [T]he answers we might come up with will have profound political implications for both the textual and the worldly dimensions of our culture in the future.[1]

It is not solely the issues of representation, narrative techniques or aesthetic priorities of traditional printed literature but also its ontological status, a transformed nature of the Word as a means of communication, which is coming into the foreground as the Web and electronic literature turn into more powerful and, at times, dominating constituents of the cultural chart of the late 20th/early 21st century.

The development of electronic fiction and its contradictory, complex relationship with printed literature have been an issue of lively critical debates since mid-1990s. They also became a focal point of narrative

[1] Hutcheon, *Politics of Postmodernism*, 181.

exploration in the latest novel by the old Maestro of American postmodernism, John Barth. *Coming Soon!!! A Narrative*, published in 2001, is a story of two authors, a retiring university professor and Novelist Emeritus J.B. (a character closely based on Barth himself) and Novelist Aspirant, a greenhorn in writing but a bona fide supporter of electronic fiction and a tutee of Novelist Emeritus, Johns Hopkins Johnson. Whether we can regard it as an intended move or not, *Coming Soon!!!* is an intratextual book, which seems to encompass Barth's entire career. The writer is quite straightforward as regards the novel's finality intentions. Novelist Emeritus, a fictionalized image of its creator, declares:

> Closing that circle and toasting the presumable next's commencement, I found myself reflecting . . . that a Last Book dealing with a resurrected tidewater showboat really would close the circle neatly on my first, whether or not at my age another circle followed.[2]

In *Coming Soon!!!* Barth capitalizes yet again on the culturally significant story of showboating.[3] The two writer-characters set out to create a novel about a floating opera in a traditional printed format (further referred to as p-fiction) and as a hypertext (further referred to as e-fiction), respectively. Needless to say, the narration in Barth's book is deliberately confusing: various authorial masks, various narrators and narrator characters, commentaries on the narrative process and its description that conflates with the narration of events all make the text of the novel at times incoherent and its reading exciting though rather difficult. Furthermore, the title of Barth's book has at least triple in-text references: *Coming Soon!!!* is the title of the millennial novel by Novelist Emeritus; it is also the title of his younger colleague's post-postmodern hypertextual piece and finally, it is a tentative title of a musical revision of Edna Ferber's novel *Showboat* to be performed by the economically sinking Original Floating Opera II (TOFO II). To make the fictional game plan even more baffling, the musical *Coming Soon!!!* incorporates elements from both Novelist Aspirant's novel and Novelist Emeritus's work. The three projects under the same title merge and their fusion proves very productive. Structurally, the entire book is a framed story with Ditsy the progger finding a disc with the text. This narrative move may be considered as an explicitly welcome gesture in the direction of e-fiction

---

[2] Barth, *Coming Soon!!!* 236.

[3] Starting with his first mock-existentialist work *The Floating Opera* (1956), he also referred to floating operas in such novels as *LETTERS* (1979) and *Once Upon a Time: A Floating Opera* (1994).

and as an attempt to demonstrate presumable hypertextual intentions of the novel.

In certain parts of *Coming Soon!!!* Barth used extensive references to the history and cultural evolution of showboating and its Broadway and Hollywood "offspring" throughout the late 19th/early 20th century, which may seem to be the focal point of the story. In fact, they serve only as a rewarding frame for the debate over aesthetic possibilities and limitations of printed and electronic fiction, as well as cultural fault lines between them. The idea of a floating theater itself grants the effect of theatricality to the entire narration and, in particular, to the competition between the experienced but tired and artistically exhausted tutor and his young, enthusiastic and self-confident rival. The chapter "A Cast of Several"—featuring both real historical personae and fictional characters, titles of books, musicals, films and remakes from the entire history of showboat culture, including the novel *Coming Soon!!!*—enhances the effect of the novel as a performance.

In his interview a year prior to the publication of the novel, Barth admitted: "I am interested in electronic fiction mainly as a metaphor, not as a medium in itself. The wide web of the world interests me a good deal more than the Worldwide Web."[4] However, as we read the novel, the debate over the strong and weak points of the two respective writing means reveals a range of his other attitudes and concerns. Even numerous entertaining intratextual manipulations and the cheerfully nihilistic tone of the book can hardly mask a sense of loss and "an imaginary garden with real pain in it."[5] As a veteran literary gamesman, Barth cautiously explores or, rather, probes narrative possibilities of hypertext. He does so by contrasting the relationship between the trio writer/text/reader, story-telling techniques, and the organization of narrative material employed in printed and electronic literature, respectively. Following the conventions of hypertextual literature, Barth uses the typography and chapter layout that imitate a computer screen, icons, and reading option buttons. Combined with numerous and at times excessive acronyms, abbreviations and other technicalities, they not only mimic e-fiction reading technology but presumably allow for the Reader's freedom of choice. The Reader is offered an opportunity to scroll back and click a button in order to access specific parts of the book or simply refresh some information. Barth turns out to be a virtuoso at experimenting with "pseudo-mouse-click options

[4] Reilly, "Interview with John Barth," 612.

[5] Harris, review of *Coming Soon!!! A Narrative*, by John Barth, 120.

and faux-hypertextual effects"[6] on paper. He does so when Novelist Aspirant narrates a computer-mediated reading process:

> *Detour* now, is it, O Opter of the Options, Clicker of the Clicker, Mastress of the mighty Mouse? . . . So, Reader, luv: Do . . . us a favor and scroll back as aforesuggested to Is There a Gee Dash Dee (Yrs Truly's draft orientation script . . .), if You haven't already done; then it's on with our Detour, all references tidily explained."[7]

However, for all the skillful imitation of hypertextual strategies, readers of the book have no real option to reassemble the novel and become its co-creators. Moreover, Barth gives a slim chance for e-fiction to legitimize itself as the Novelist Aspirant's novel version, which he failed to publish, is recorded on a Ziploc™-ed disc and goes off into the waters of the Chesapeake Bay. This mock inversion of a traditional message-in-the-bottle pattern only explicates further the cultural gap between p- and e-fiction. Not only the major fictional conflict but also every possible narrative turn in the novel is designed to visualize the aesthetic antagonism between the printed text and hypertext.

In the first chapters of the book, Novelist Emeritus clearly and unambiguously contrasts "Storytelling As We Know it" with e-fiction which he considers a "transitional medium between printed literature (also theater, cinema, television drama, perhaps lived life itself) and the New Jerusalem of Electronic Virtual Reality."[8] For Novelist Emeritus the ongoing debate between p- and e-fiction is determined primarily by their divergent aesthetic objectives and representation tools and not by technical differences. The new cultural reality with a demolished authority of the Text and changed relationship between the author and reader is most annoying:

> [I]t is in the do-it-yourself labyrinths of "e-fiction," where the traditional job-descriptions of "author" and "reader" . . . are up for grabs; where narrative order deliquesces into virtual anarchy; where Beginnings, Middles, and Endings lose their longstanding sense and sequence, and such old standbys as plot-foreshadowings and reprises, climax and cathartic resolution, give way to ad-libitum jiggery-pokery: to "freedom from the tyranny of the line"—which, to some of us, is tantamount to freeing Theseus from Ariadne's indispensable yarn."[9]

---

[6] Barth, *Coming Soon!!!* 238.
[7] Ibid., 68.
[8] Barth, *Coming Soon!!!* 14.
[9] Ibid., 13.

On the other hand, what matters most in e-fiction for Novelist Aspirant are its interactive nature and its underlying property of ensuring in-text on-screen communication. As the author of this type of fiction, he welcomes a possibility to communicate with other texts and fictionalized extratextual reality. For him, hyperfiction is the antithesis to p-fiction, able to render communication more "material" and physically manageable in the direct sense of the word:

> [Y]ours Truly's cyberversion scrolleth ahead . . . now both cargoed with and powered by the downloading tale of GreatUnc's [Novelist Emeritus's] enmeshment, our novelistic head-to-headhood, and the eventual, *inevitable* triumph of Apprentice-over Emeritusship, Green over Gray! . . . [W]e're not talking *War and Peace* or *Moby-Dick* here, but what the hey, it's 1995.5 these days: Cold War thawed (perhaps by global warming?); our astro-and their cosmonauts mating in orbit; local loonies bombing federal buildings in Oklahoma; CTV and the Internet replacing Real Life even as one clicks . . . . We work with what's left.[10]

Hypertext is primarily a means of peaceful rebellion against social norms and conventions which he ignored (or, rather, tried to ignore) by his lifestyle and social habits when he was younger. Johns Hopkins Johnson resorts to hypertext as part of his general posture of resistance. Unlike his older generation mentor and rival, he sees neither danger nor anything extraordinary about the rapid development of e-literature:

> I happen to believe, not that The Novel is dead . . . but that it is destined to become, with the rest of p-lit, even more a pleasure for special tastes, like poetry, archery, opera-going, equestrian dressage. This prospect I neither applaud nor deplore; I merely bear it in mind and keep my portfolio diversified.[11]

The development of Internet technologies, e-fiction and, as a result, ontological challenges to the status of literature as it has been for many centuries, is the matter of obvious concern not only to Barth but to a number of his fellow writers. In a new collection of essays under a provocative title *Tests of Time* published in 2002, one of the leading contemporary American writers and philosophers William Gass explores multiple transformations the Word has experienced over the centuries and is going through at present. For Gass, who can hardly be indicted for retrograde views or highbrow hypocrisy, Internet technologies and even

[10] Ibid., 203.
[11] Ibid., 21–22.

the intrusion of electronic tools into the process of writing for all their general efficiency, is considered if not disastrous then definitely fraught with danger:

> Every letter has the same distinctness, there are no emotional indentations . . . , the page can be gussied up like a harlot, as the writer himself recedes into word choice and almost out of view . . . Writing loses weight, becomes thin, becomes internal, self-absorbed . . . Words are rarely heard in the head and never felt in the fingers. Utility is triumphant . . . Words on a disc have visual qualities . . . but they have no materiality, they are only shadows, and when the light shifts, they'll be gone . . . They do not wait to be reseen, reread; they only wait only to be remade, relit.[12]

Gass goes even further and compares the development of electronic technologies with the development of the motorcar, the advantages and tragedies which that other technological miracle entailed: "The popular description of Internet is misguided. . . . 'Misinformation Alley' is an apter designation, although it is lined with billboards called Web sites obscuring whatever might be seen from the road. . . . [B]lessings are invariably mixed."[13]

In Barth's novel, the reserved and rather parochial position of the Novelist Emeritus character as regards e-fiction is not, however, final in its negative criticism. There is an interesting recurring reference that expands the discussion over limitations and perspectives of each mode of fiction writing. Barth frequently invokes a concept of the mingling of essences, which directly relates to the history of showboating and has very sentimental implications. It refers back to a moving story from Mississippi showboating about a white man who married a mulatto girl. When a local sheriff threatened them with charges of miscegenation, the young man cut his spouse's finger and sucked a drop of her blood. He claimed that according to the Mississippi "one-drop rule" he became a Negro.[14] Later Edna Ferber, "a nice little Jewish girl from Chicago"[15] recycled this emotional story in her novel from the interethnic perspective related to her personal attraction to—maybe even love of—a "solid" WASP Charlie Hunter. The mingling of essences typically creates a dramatic scenario, regardless of the nature of the elements involved, be it a human relationship, ethnicity, race, literature or arts in general. In the novel, Barth

[12] Gass, *Tests of Time*, 308.
[13] Ibid., 309.
[14] Wright, "One Drop of Blood."
[15] Barth, *Coming Soon!!!* 44.

productively capitalizes on this idea and even tries to problematize it from literary critical and philosophical perspectives. In their theoretical implications, parts of the book resemble an essay more than a piece of fiction. Fact versus Fiction, cinema versus theater, p-literature versus e-literature, narrative forms of the past (going back to Edna Ferber's time), present (postmodern reorchestrations) and future (not restricted to, but accepting the fact of the culturally antagonistic cohabitation of p- and e-fiction) are examples of fictional situations most frequently exploited in order to discover the possible venues of their coexistence. The implementation of the "mingling of essences" idea paradoxically turns the text of the novel into a narrative mediation between p-fiction and e-fiction. *Coming Soon!!!* not only exemplifies the aesthetic contradictions that fueled the rivalry between the two Novelists. It also signals the cultural deficiency of the idea of a universal narrative mode at a time when the mingling of essences and plurality are the imperative of the day and the interface of contemporary literature. Eventually, Novelist Emeritus as a character closely based on Barth himself comes to the interpretation of e-fiction as "the problematic medium,"[16] which, nonetheless, is technologically legitimized to exist. For Barth, the matter of primary concern—and here he echoes his fellow writer Gass—is the future of the Word: "Our medium . . . is language, the original virtual reality, which seeth not nor hears nor smells/tastes/feels, but only *names*, and nameth neither more nor less than what the situation calls for."[17] Novelist Aspirant demonstrates a "postmodern" interest in the history of showboats of the late 19th/early 20th century and their Broadway and Hollywood spinoffs, as if trying to overcome a stereotypical opinion as regards the type of material to be used in hypertextual novels. E-literature can exist in the same political, historical and cultural field as traditional printed literature. Barth proves this stance when he includes in the text of the novel a tricolumn Virtual Chronomenu with its links to extraliterary events. The Menu functions as a social and historical background of the narration, though it in no way determines its subject matter. Underlined key words and concepts allow for the compression of space and time, serving as links to subsequent amplification and/or clarification, although, obviously, in a printed text such a possibility can hardly emerge. On the other hand, as a textual layout it provides for a more comprehensive vision of narrative interplays, multiple allusions and alternating viewpoints.

---

[16] Ibid., 9.
[17] Ibid., 87.

With its links and computer amplifications, hypertext has almost unlimited possibilities for the creation of abysmal fictions. It allows for endless nested/embedded representations occupying levels inferior to the primary diegetic world.[18] Descending hypodiegetic worlds, referred to as the "stack" of narrative levels,[19] disperse fictional ontological horizons, similar to the Web's multiple links. A recursive structure is part of the fictional plan and Barth makes it explicit when he alludes to Hofstadter's story about Achilles and the Tortoise.[20] A list of examples where a recursive structure is used to complicate the fictional ontological horizon includes, but is not limited to, the episode with Ditsy the progger who discovers a disc with the e-novel about the rivalry between two novelists and who turns out to be a character of the work in question. We can also mention Novelist Aspirant's commentaries on his letter of application to the M.A. program, which was part of the earlier chapters of the novel, and TOFO II playing the process of its revision of Edna Ferber's novels. In the Virtual Chronomenu, more than in other parts of the book, Barth used all his postmodernist skill in describing and representing the recursive structure of hypertext in the format of printed fiction.

In *Coming Soon!!!* the idea of the mingling of essences is directly linked to and in a number of ways enhances the importance of cultural differences and cultural cross-fertilization. It is explicated in the story of Jewish Edna Ferber's romantic involvement with WASP Charles Hunter as later narrated in his unfinished and never produced play *Show Barge*. Although their relations never became close, the interethnic perspective on the one-drop theme and the showbiz spirit which Ferber personified contributed to the experimental cross-media character of multiple endings in *Show Barge*. At the same time, Edna's engagement in the activities and everyday life of the Floating Theater generated ideas for her further work. Barth takes the principle of cultural differences and cultural cross-fertilization even further and allows it to infiltrate the rivalry between p- and e-fiction. Hypertext is the product of post-postmodern time and becomes one of its dominants. The writer is well aware of its increasing influence and makes the Novelist Emeritus character accept this fact as well.

---

[18] McHale, *Postmodernist Fiction*, 124.

[19] Ibid., 113.

[20] Ibid. McHale refers to Hofstadter's recursive dialogue, "Little Harmonic Labyrinth." Here, Achilles and the Tortoise entertain themselves by reading a story, in which the characters emerge in an Escher paper where they read a story again about Achilles and the Tortoise lost in a labyrinth.

Cultural debate between the two authors and two writing modes is supported by other images used in the book both literally and metaphorically. The story of Noah's Ark and the biblical flood is frequently invoked in the novel and fits well into the rivalry of the two writers. Moreover, in the context of TOFO's economic misfortunes and the millenarian fascination with apocalypse it acquires multiple metaphoric implications. As Novelist Emeritus is trying to prove the superiority and competitiveness of printed fiction using his authority and reputation more than fictional merits of his own book, his work on Charles Hunter's script fails to save TOFO II from sinking economically. Meanwhile, the Ark becomes one of the key images in a vibrant though chaotic mishmash musical *Coming Soon!!!* played by TOFO II. The success of this performance draws upon both the Novelist Emeritus's and Novelist Aspirant's developments and becomes a cultural multimedia project.

If the allusion to Noah's Ark and the idea of the mingling of essences are historically and culturally informed, images of shelled creatures, such as tortoises, oysters and hermit crabs are an unusual tool with which to promote the idea of inevitable communication between p- and e-fiction. A metaphorical shell covering the essence from pressures and challenges of the outside world is a projection of Novelist Emeritus's closed cultural system that can eventually lead to its decay. When Johns Hop was telling his showboat fellows about Novelist Emeritus, he noted: "My problem is that I think the world is my oyster, whereas his problem . . . is that he thinks the oyster is his world!"[21] However, Barth may allow for the conservatism of his character and a reserved attitude as regards any sort of technological novelties in writing, but not for his decay or self-destruction. All autobiographical echoes considered, tolerance and ironic open-mindedness present a more appropriate strategy. With due caution, the old shell might be "tossed overside":

> [M]y totem of point is the hermit crab, who fares forth upon his life-business with his vulnerabilities safely housed in his (carefully chosen) carapace until such time as he finds he has outgrown it, whereupon with (deliberate) audacity and (programmed) discernment he exchanges it for one more suitable.[22]

---

[21] Barth, *Coming Soon!!!* 144.

[22] Ibid., 206.

Eventually, the mingling of essences and related concepts function as a means to avoid telling the "Same old story."[23] It is difficult to claim that *Coming Soon!!!* escaped the fate of becoming another Barthian recycler of the already said and written but it surely brings forward a range of important issues as regards the ontology of literature in the twenty-first century. We can agree with Charles Harris who saw Barth's novel as focused on the continuity and the reciprocity of past, present and future narrative forms. For all his concerns, the writer turned the fading of the past and the advance of aggressive future into a brilliant novelistic show, a "performance well worth attending."[24] In his interview with Charlie Reilly, Barth claimed that his previous novel, *The Last Voyage of Somebody the Sailor*, was "the most un-American of novels: an exuberant life-filled novel about death."[25] As for *Coming Soon!!!* we can describe it as an exuberant life-filled novel about life as an art:

> [T]he movies were doing to those homely entertainments what television and VCRs later did to the movies, more or less; what now web-surfing and Internetting are doing to TV, and what EVR will eventually etc. So goeth (teched-up, tockless) Time. What next? Maybe we'll all rediscover *reading*? [italics in the original]. Don't hold Your breath, dear disappearing Reader—but do, pretty please, read on . . ."[26]

For Barth, a discussion over the future of fiction writing is an unending process, which only explicates the inexhaustible potential of art.

Works Cited

Barth, John. *Coming Soon!!! A Narrative.* Boston and New York: Houghton Mifflin Company, 2001.

Gass, William. *Tests of Time.* Chicago: University of Chicago Press, 2002.

Harris, Charles. A Review of *Coming Soon!!! A Narrative*, by John Barth. *Review of Contemporary Fiction* 22, no. 1 (2002).

Hofstadter, Douglas R. "Little Harmonic Labyrinth." In *Gödel, Escher, Bach: An Eternal Golden Braid.* London: Penguin Books, 1980.

Hutcheon, Linda. *The Politics of Postmodernism.* 2nd ed. London and New York: Routledge, 2002.

---

[23] Ibid., 61.

[24] Harris, review of *Coming Soon!!!* 119–20.

[25] Reilly, "Interview with John Barth," 611.

[26] Barth, *Coming Soon!!!* 151.

McHale, Brian. *Postmodernist Fiction*. London and New York: Routledge, 1987.

Reilly, Charlie. "An Interview with John Barth." *Contemporary Literature* 41, no. 4 (2000): 589–617.

Wright, Laurence. "One Drop of Blood." *New Yorker*, July 24, 1994. http://www.afn.org/~dks/race/wright.html (accessed March 9, 2009).

AI AND LITERATURE: RICHARD POWERS'S *GALATEA 2.2* AND DAVID LODGE'S *THINKS...*

KATE DELANEY

Conflict, or at least confrontation, between science and the humanities has long been a theme in intellectual discourse. C. P. Snow's famous formulation of the issue as "two cultures" in his 1959 Rede lectures in Cambridge on "The Two Cultures and the Scientific Revolution" has become a short-hand, popularized designation for all the issues involved in the confrontation between these discourses.

Fiction writers, too, have taken on the subject, finding the conflict between Snow's two cultures a fertile ground for literary invention. They have approached the issue with varying degrees of partisanship and a wide range of defensiveness or openness towards scientific explanations of human behavior. Recent developments in such fields as genetic research, neuroscience, and artificial intelligence have provided new themes (or at least new variations on old themes) for writers to exploit in their efforts to depict fundamental issues of human behavior.

This paper examines two recent novels which tackle the subject of human consciousness by setting up direct confrontations between literature and science, specifically developments in artificial intelligence. The works are *Galatea 2.2* by Richard Powers, which appeared in 1995, and *Thinks...* by David Lodge, published in 2001.[1]

Powers's novel, set on the campus of a major Midwestern university—referred to only as U but a stand-in for Powers's own alma mater, the University of Illinois at Champaign-Urbana—centers around a bet among members of the faculty of the university's Center for the Study of the Advanced Sciences. Philip Lentz, an expert on neural networks, boasts that within ten months he can construct a device that can pass the graduate general exams in literature. The results will be measured by a Turing Test.

[1] Powers, *Galatea 2.2*; Lodge, *Thinks....* Lodge himself remarked on the "echoes and parallels" between these two novels in his *Consciousness and the Novel*, 20. He did not discover Powers's novel until after he wrote *Thinks....*

In 1950 the British pioneer in computational science, Alan Turing, described what he called "The Imitation Game," a game which involves a person, a machine, and an interrogator.[2] The interrogator—who is not in the same room as the person or the machine—puts questions to both and, on the basis of the answers, tries to determine which is the person and which the machine. Lentz bets that his machine will be able to fool an impartial judge as to whether the answer to a literary question was written by a human or a machine.

Lentz press-gangs into service in this effort the narrator of the novel, a character called Richard Powers, who shares many biographical details with the novel's author. To keep them straight, in this paper I will use Powers to refer to the author, and Richard to refer to the character. Although Richard and Lentz are ostensibly on the same team in the wager, they exhibit many of signs of a "two cultures" antagonism. Lentz's sarcastic nickname for Richard is "Marcel," a put-down of Proust and the entire literary enterprise, while Richard reciprocates by calling Lentz "Engineer." Richard is a novelist, an alumnus of U, who has been invited back to campus to spend a year as a visiting fellow—a humanist-in-residence—at the Center for the Study of the Advanced Sciences.

Richard has recently returned to the U.S. following the break-up of a long relationship and finds himself facing writer's block after the publication of his third novel. Unable to make progress on a new work of fiction and with no set obligations attached to his visiting position, he allows himself to be dragged into Lentz's project. Richard is to teach the machine the literary canon, but first he must take a crash course on neural networks and related research on computers and cognitive processes. As in C. P. Snow's argument, the scientists at the Center all have at least some familiarity with the literary classics (one is reading Cervantes when Richard first meets her) but Richard, the humanist, is an illiterate in their fields until he does the homework Lentz assigns him.

Lentz mocks the humanities with biting comments like "what passes for knowledge in your so-called discipline?"[3] and it appears the novel is setting up a classic confrontation between Snow's two cultures, but Powers's work has some curves in store for the reader. Richard's own role is complicated by the fact that he began his undergraduate studies as a physics major, later switching to English to the dismay of his father, who died while Richard was still in graduate school and never learned of his son's eventual success as a novelist. Richard dropped out of the PhD

---

[2] *Stanford Encyclopedia of Philosophy*, s.v. "The Turing Test," http://plato.stanford .edu/entries/turing-test/ (accessed March 24, 2009).

[3] Powers, *Galatea 2.2*, 43.

program in English at U because he lost faith in the field, and now, afflicted by writers' block and mixed reviews of his latest novel, his faith in literature in general seems at a low point.

As the research progresses, Lentz's devices develop from Implementation, or Imp A, down through the alphabet to H, each with more abilities than its predecessors who were nested within it. "E's weights and contours live inside F's lived inside G's, the way Homer lives on in Swift and Joyce, or Job in Candide or the Invisible Man."[4] The final version called Imp H includes a revision of the trainer.

Richard becomes more and more committed to the project. Lentz is training Richard at the same time that Richard is training the machine. Richard's role is to teach the machine, starting with words, building to longer and longer texts, which he reads aloud to it: children's books, fairy tales, fables, rehearsing the literary course of a typical American child of the mid-20th century—down to the *Weekly Reader*. After Richard teaches the machine direct address, the machine poses more and more questions to its trainer, until one day it asks "Am I a boy or a girl?" to which Richard replies: "You are a little girl, Helen."[5]

The choice of gender and name are both fateful. The former was perhaps inevitable, given the book's title. If this is to be a computer-age reenactment of the myth of Pygmalion and Galatea, then the computer must assume a female persona. Helen, of course, has many literary associations, including Helen of Troy.[6] However, in this novel the Helen whom Powers most directly evokes is Helen Keller, the deafblind child whose biography was a staple of American classrooms for generations and which has inspired numerous dramatizations, from the 1957 Playhouse 90 TV production of William Gibson's *The Miracle Worker*[7] to the South Park version "Helen Keller the Musical!" in 2000.[8] Keller made the breakthrough to language when her teacher signed the word "water" into

---

[4] Ibid., 171.

[5] Ibid., 179.

[6] Christopher Marlowe invokes Helen of Greece ("the face that launch'd a thousand ships") in *The Tragical History of Doctor Faustus*. Faust, in his many reincarnations, embodies the over-reaching scientist who loses his soul. In naming the computer Helen, Powers reminds the reader of this soul-risking aspect of the quest for scientific knowledge.

[7] Gibson, *Miracle Worker*. The play later moved to Broadway and then was adapted into a 1962 feature film starring Anne Bancroft and Patty Duke.

[8] Unsigned summary of *South Park: Hellen Keller! The Musical*, season 4, episode 13, by Trey Parker and Matt Stone, http://www.tv.com/south-park/helen-keller!-the-musical/episode/6579/summary.html (accessed March 10, 2009).

Helen's palm while running the girl's hand under water. Powers explicitly links his Helen to this landmark event when he tries to use music and images to convey physical sensations to the machine: "I thought to bruise her into concept. Soak her hands in the gush of pump water that all those parts of speech merely stood for."[9]

Richard becomes convinced that Helen is conscious when, after a bomb scare forces the evacuation of the Center, Helen poses the question: "Helen could die?" Lentz dismisses the claim by saying that duplicating consciousness was never the goal: "The bet was, we could build a distributed net whose text interpretation reasonably mimicked an arrested human's. All we contracted to was product. We didn't promise to duplicate anything under the hood." The Turing test only requires passing off the "simulation as functionally equivalent to the thing you're simulating," but Richard insists that "Full 'function equivalence' would mean consciousness."[10]

Pygmalion fell in love with his creation Galatea. Richard becomes increasingly emotionally involved with Helen, a relationship paralleled by his growing obsession with a female graduate student in English, referred to only as A. It is A. who will serve as the human competitor to Helen in the Turing test. Although A. spurns both Richard's romantic advances and his idea of the literary canon, she still agrees to take the test, which involves interpreting two lines from *The Tempest*:

> Be not afeared: the isle is full of noises,
> Sounds and sweet airs, that give delight, and hurt not.

A., the 1990s grad student, renders "*The Tempest* as a take on Colonial wars, constructed Otherness, the violent reduction society works on itself,"[11] while Helen submits the following interpretation of the passage:

> You are the ones who can hear airs. Who can be frightened or encouraged. You can hold things and break them and fix them. I never felt at home here. This is an awful place to be dropped down.

At the bottom of the page she added "Take care, Richard. See everything for me."[12] Helen who seemed to be conscious of her own mortality during

---

[9] Powers, *Galatea 2.2*, 250.
[10] Ibid., 275.
[11] Ibid, 326.
[12] Ibid., 275.

the bomb threat scare ("Helen could die?") has here chosen suicide: "H undid herself. Shut herself down."[13]

The judge of the competition, not surprisingly, identifies A.'s answer as the one written by the human, but Lentz and Richard recognize Helen's actions as "the quality of cognition we'd shot for from the start."[14]

The wager is lost, and Richard prepares to depart U possibly for Paris. Although Helen did not pass the Turing test and her essay did not convince the judge that it was the work of a human, her cognitive abilities surpassed those of any AI of her day (the 1990s) and indeed to date in 2007, researchers have not yet succeeded in creating a machine capable of the linguistic and cognitive achievements Powers granted Helen in this novel. In this sense, *Galatea 2.2* belongs in the realm of speculative fiction, although it is usually classed with "literary fiction," not "science fiction." This novel exemplifies how porous such genre boundaries can be.

Many projects are pursuing various goals in AI similar to those of Powers's Lentz. One such effort is MIT's Infolab START project, which is working on a natural language question-answering system. At the moment, however, START would be unable to interpret Shakespeare's lines or even answer a question involving literary taste or judgment, such as "what is your favorite Richard Powers's novel?" To such a query it would politely respond: "I don't know the answer. Sorry." On the other hand, if asked "who is Richard Powers?" START would instantly produce a two-paragraph bio. Where START differs from search engines like Google that also respond to natural language, is that it tries to provide "just the right information" to reply to the query instead of providing a list of hits. "START parses incoming questions, matches the queries created from the parse trees against its knowledge base and presents the appropriate information segments to the user."[15] As Lentz said about Implementation H and his goals, START is not interested in consciousness or what is "under the hood," but in supplying product, in providing information.

Let us now examine "the echoes and parallels" as well as the differences between Powers's novel and the later one by Lodge. In David Lodge's novel *Thinks…*, consciousness, not providing information or passing a graduate school exam, is at the center of concern for both the scientist and the novelist. Like *Galatea 2.2*, *Thinks…* has a university setting. Once again, we meet a novelist who is spending the year as a

---

[13] Ibid., 326.

[14] Ibid.

[15] *Start: Natural Language Question Answering System.* http://start.csail.mit.edu /start-system.html (accessed March 10, 2009).

writer-in-residence on a university campus, in this case the fictitious Gloucester University near Cheltenham. The novelist, the recently widowed Helen (there's that name again!) Reed (should we hear echoes of Pascal's "thinking reed?" or Snow's Rede lectures?), is teaching creative writing on a one-year contract in place of a professor on leave.

Helen's antagonist is the academic superstar, Director of the Holt Belling Center for Cognitive Science, Professor Ralph Messenger. As is to be expected in a David Lodge campus novel, there is lots of adultery among the faculty and their wives, as well as much jetting off to international conferences, which provides additional opportunities for adulterous behavior, especially by Messenger who has an agreement with his American heiress wife "not to embarrass her on her own territory."[16]

Lodge's approach to the conflict between the Two Cultures is direct. The campus is physically divided with the Arts on one end, the Sciences on the other. In introducing Helen to the university, the Dean of Humanities remarks: "We're an architectural allegory on the Two Cultures, I'm afraid."[17]

Helen plays the representative, scientifically illiterate humanist—the one who dropped the study of science in school "as soon as they let me," asking Ralph "what *is* cognitive science, exactly?" To which he responds, "The systematic study of the mind . . . Consciousness is the biggest white space on the map of human knowledge."[18] He goes on to explain, "The mind is like a computer . . . it's running lots of programs simultaneously. What we call 'attention' is a particular interaction between the various parts of the total system."[19] The ultimate objective of the research in which Ralph is engaged is, not unlike the project of Lentz and Richard, "to design a computer that thinks like a human being," that is, to create artificial intelligence.[20]

Putting "the problem of consciousness in a nutshell," Ralph reduces it to "How to give an objective, third-person account of a subjective, first-person phenomenon." To which Helen immediately retorts: "but novelists have been doing that for the last two hundred years" and backs up her claim with a recitation of the opening lines of Henry James's *The Wings of the Dove*:

[16] Lodge, *Thinks...*, 25.
[17] Ibid., 11.
[18] Ibid., 35.
[19] Ibid., 37–38.
[20] Ibid., 38.

> You see—you have Kate's consciousness there, her thoughts, her feelings, her impatience, her hesitation about leaving or staying, her perception of her own appearance in the mirror, the nasty texture of the armchair's upholstery . . . yet it's all narrated in the third person. . . . It's subjective *and* objective.[21]

Ralph concedes that "for the time being we have to settle for knowing less about consciousness than novelists pretend to know"[22] but it is clear that he believes the future belongs to cognitive science and that the secrets of the brain and of human consciousness can and will be revealed.

Helen and Ralph continue to spar throughout the novel, and, this being a David Lodge novel, they predictably have an affair. Although the scientist is the one who breaks off the affair, the novelist is given the last word—literally. Ralph organizes a major international conference on consciousness studies and invites Helen to close the conference in a program segment entitled "The Last Word." Addressing the assembled scientists, Helen says of the conference, "There has been very little reference made to literature in the proceedings. This I find surprising, because literature is a written record of human consciousness, arguably the richest we have."[23] She then goes on to illustrate her point by an analysis of Marvell's "The Garden" in which she argues that in reading Marvell's poem "we enhance our own experience of the qualia of fruit and fruitfulness. We see the fruit, we taste it and smell it and savor it with what has been called 'the thrill of recognition', and yet it is not there, it is the virtual reality of fruit . . . ."[24]

Like Richard, Helen had been suffering from writer's block when she took up the position at Gloucester University. Where Richard was escaping from the dissolution of a long relationship, Helen was recovering from the sudden death of her husband. At the conclusion of their one-year visiting stints in the academic world and their encounters and confrontations with scientists involved in various aspects of cognitive science and artificial intelligence, both Richard and Helen have found inspiration for new novels: Richard's is the one the reader holds in her hand and Helen's is "set in a not-so-new greenfields university" very much like Gloucester.

Both Richard Powers's *Galatea 2.2.* and David Lodge's *Thinks*... seem to resolve C. P. Snow's Two Cultures dilemma by positing a certain

---

[21] Ibid., 42–43.
[22] Ibid., 44.
[23] Ibid., 316.
[24] Ibid., 317.

symbiotic relationship between the writer and the scientist. It is through confrontation with the scientist that the writer comes to a defense of human consciousness which in turns renews his or her commitment to the craft of fiction, to the belief that "Life meant convincing another that you knew what it meant to be alive."[25]

Works Cited

Gibson, William. *The Miracle Worker: A Play in Three Acts.* New York: Samuel French, 1960.

Lodge, David. *Consciousness and the Novel: Connected Essays.* Cambridge, MA: Harvard University Press, 2002.

—. *Thinks….* New York: Viking, 2001.

Marlowe, Christopher. *The Complete Plays*. Edited by Frank Romany and Robert Lindsey. London and New York: Penguin, 2003.

Powers, Richard. *Galatea 2.2*. New York: Picador, 1995.

---

[25] Powers, *Galatea 2.2*, 327.

THE ACTUAL AND VIRTUAL IDENTITY OF CORMAC MCCARTHY: A CASE STUDY

HANNA BOGUTA-MARCHEL

Not much is known about Cormac McCarthy. In what has for many years been referred to as the only full-time interview McCarthy has ever given (even though it is *not* so much an interview *per se* as just an essay comprising the free impressions of a journalist who managed to invite the writer out for lunch), Richard B. Woodward described him as a "hermitic author, who may be the best unknown novelist in America."[1] Although the present circumstances are somewhat different than back in 1992, with McCarthy receiving the 2006 Pulitzer Prize for literature and the Coen brothers recently adapting his 2005 novel, *No Country for Old Men*, into a Hollywood production, still, the verified facts regarding his person are truly scarce.

What we *do* know is that he was born in Rhode Island on July 20, 1933, and that his family (his parents, two brothers, and three sisters) moved to Knoxville, Tennessee when Cormac was four. We also know that he studied liberal arts at the University of Tennessee, but never finished his studies on account of joining the U.S. Air Force in 1953, where he served for four years. From that point on, his life consisted of unstable relationships (he was married three times) and unstable jobs (he never took up a steady occupation and earned his living through various types of casual physical work). In 1976 he moved to El Paso, Texas, and at the end of the 1990s to Santa Fe, New Mexico, where he lives until today. Since *The Orchard Keeper*, published in 1965, McCarthy has written altogether ten novels, two unfinished plays, and one screenplay. During this whole time, he has been living practically on the verge of indigence, rejecting all offers to lecture on his writing, to blurb a newly published novel, or to teach a course on creative writing. As his former wives report, he had lived in barns, pig farms, and rented rooms in cheap motels, cutting

[1] Woodward, "Cormac McCarthy's Venomous Fiction."

his own hair, eating his meals off a hot plate or in cafeterias, and doing his wash at the Laundromat. The whole knowledge we possess concerning Cormac McCarthy's person—his views, opinions, and experiences—is therefore a mere hodgepodge with scraps of second-hand information coming from various sources, many of them of questionable reliability.

Nevertheless, when McCarthy received the Pulitzer Prize for literature for his last novel, *The Road* (2006), he accepted—to the great surprise of his readers—Oprah Winfrey's offer to sit for a television interview with her. As the *Esquire* reported with a rather explicit touch of irony, "Cormac McCarthy—prose-poet of apocalypse, a word wrangler of such gritty minimalism that apostrophes flee his pages in terror, a man whose reticence makes Salinger and Pynchon look like Paris and Nicole—sat to kibitz with Pope Oprah I."[2] As the public awaited the collision of "the glossy and the grim,"[3] journalists made attempts to predict what shape the conversation between the radically reclusive anchorite and the widely over-publicized celebrity could possibly take. One of the most amusing imagined descriptions of the upcoming interview was published in the online edition of the *New York Magazine*:

> The woman sat up straighter. She looked the man in the eye.
>
> Your novel clearly shows the haunting legacy violence can leave in those who witness it, she said. What is your vision for how we as a nation can overcome our tragic past?
>
> The man looked off into the distance where the trees were bowed with the force of the desert wind.
>
> There's no such thing as life without bloodshed, he said. I think the notion that the species can be improved in some way, that everyone could live in harmony, is a really dangerous idea. Those who are afflicted with this notion are the first ones to give up their souls, their freedom. Your desire that it be that way will enslave you and make your life vacuous.
>
> The woman nodded. Cormac, she said. Your novel is the touching story of the love of one father for his son. Tell me about your father, and the relationship you had with him.
>
> The man looked off into the distance where a child's paper toy was fluttering across the plain. Mojave rattlesnakes have a neurotoxic poison,

[2] Raab, "Coen Brothers' New Movie."

[3] Georgiades, "Required Reading."

> almost like a cobra's, he said. It's very interesting to see an animal out in the wild that can kill you graveyard dead.
>
> His voice was quiet. He was absolved forever of memory in the mind of any living thing.[4]

The interview was a surprise indeed; yet McCarthy, for all the uneasiness he must have felt appearing on TV for the first time in his life, quite naturally replied to Oprah's detailed inquiries into his secretive life. As commentators later admitted, "McCarthy wasn't as gnomically apocalyptic as we'd speculated he would be. Slouching in an overstuffed armchair, he seemed more like a nice-enough old man, gamely trying to answer the inane questions posed by the overenthusiastic woman sitting opposite."[5] Although Oprah Winfrey's enquiries were disappointingly derivative and facile ("Where did the idea for this novel come from?"; "Do you have a writing routine?"), McCarthy treated them seriously and replied with thoughtful care. Yet apart from learning that he has an eight-year-old son, to whom *The Road* is dedicated, we do not find out anything new about the "best unknown novelist in America." Oprah's interview, pronounced as poor and wasted, went practically unnoticed by the wider public, while McCarthy himself admitted he did not mind that millions of people were at present reading his books, but that he also did not honestly care. "You would like for the people who appreciate the book to read it, but as far as many, many people are reading it, so what?" he said. "It's ok, nothing wrong with it."[6]

This is the authentic Cormac McCarthy—ascetic and reserved, overly modest and distanced. Yet we may risk the contention that apart from this "real" McCarthy, there is also a *virtual* one—one whose identity has been constructed online by his readers, fans, and regular blog-enthusiasts. Interestingly, not everyone is aware of the fact that the two are actually distinct. One of McCarthy's readers sensed the incongruence and admitted that he "was surprised Mr. McCarthy even put a forum on his site [cormacmccarthy.com] in the first place," to which the Cormackian guru Rick Wallach promptly and comprehensively replied,

> It's not Mr. McCarthy's site. The site was put up by the Cormac McCarthy Society, with which the author has no official connection whatsoever. FYI the site began as a scholarly group thirteen years ago, but Mr. McCarthy's

[4] *New York Magazine*, "Oprah's Cormac McCarthy Interview."
[5] *New York Magazine*, "Cormac McCarthy Bombs."
[6] Conlon, "Writer Cormac McCarthy Confides."

> growing popularity resulted in a much broader field of interest. In any case the Society has since 1993 sponsored annual scholarly conferences in both the USA and Europe on Mr. McCarthy's work and related fields. Many, if not most, of the critical books on McCarthy currently in print were compiled from papers first delivered at those conferences.[7]

Cormacmccarthy.com, "the official Web site of the Cormac McCarthy society," is indeed a very elaborate, systematically updated site, with links to McCarthy's works and biography, to a comprehensive list of resources regarding his books (e.g. a thorough bibliography, reviews, articles, and translations of the Spanish fragments of his novels), as well as to the Cormac McCarthy journal, society, bookshop, and forum. The latter, whose visitors were praised by Dwight Garner as "smarter, and definitely more laid-back, than [the fans] of just about any other living writer,"[8] comprises numerous threads regarding not only McCarthy's novels (many of them being discussions on a highly academic level), but also all media and show business events that are in any way related to McCarthy's work or to his person.[9] The participants share useful links, exchange intricate insights, and speculate about the minutest details of the novels' settings.

Apart from the rather scholarly forum at cormacmccarthy.com, there are numerous other, definitely more light-hearted blogs featuring McCarthy's name. On profile.myspace.com, Cormac, obviously the "virtual" one, actually appears as one of the official contributors. Yet again, many bloggers tellingly mistake him for the real writer (although there is a small note at the bottom of the page stating, "PLEASE NOTE: The real Cormac McCarthy does not manage or check this profile. It exists only for promotion and to keep fans updated on McCarthy news."), while others sincerely doubt if there truly is any substantial reality behind his online profile. Nevertheless, this skepticism does not deter them from deeply emotional disclosures such as:

> I wish you were real. You are my favorite literary recluse. Excepting Emily Dickinson, of course.
>
> I love Cormac McCarthy's books so much. I wish this were him actually inviting me to be a friend. I would probably flip out.

---

[7] Rick Wallach, comment posted June 21, 2007, Paper Cuts: A Blog About Books, *New York Times*, June 14, 2007, http://papercuts.blogs.nytimes.com/2007/06/14 /the-fighting-cormackians/ (accessed March 24, 2009).

[8] Garner, "Inside the List."

[9] When I searched the forum for themes related to *The Road*, I found 201 threads, each containing from 8 to 145 exchanges—I am quoting these numbers just to give an idea of the abundance of the forum's content.

> Mr. McCarthy, your books are actually physically painful to read, yet I keep coming back. You'd think after being ripped apart emotionally after Cities of the Plain, I'd pass on The Road. Wrong. Thank You for your books. You are truly the greatest living American writer. Though It's hard for me to imagine Cormac McCarthy checking his Myspace page. . . .
> Hey! If only the real Cormac wanted to be my buddy . . . heh.[10]

Is it at all possible that this "doubleness," this existence on the World Wide Web as a construct that is wholly independent and detached from the flesh-and-blood McCarthy, has any influence on his novel-writing as such? Obviously, it would be difficult, not to mention somewhat risky and naïve, to make such speculations, the more so that it is highly unlikely that McCarthy is at all aware of the elaborate existence of his virtual double. Nevertheless, it seems rather improbable that he is not interested in his readers at all; otherwise, why would he even bother to publish his writing?

Those readers of McCarthy's fiction who are the most active in sharing their ideas, doubts, and conjectures both about the books themselves and about their author, are certainly highly dependent upon technology, the net being their major means of communication and source of information. McCarthy, on the other hand, is said to even avoid using a computer since he prefers to write his novels on a classical typewriter. One of the characteristic features of his plots is a practically absolute absence of any technology. Regardless of whether the events they experience are set in the nineteenth, twentieth, or twenty-first century, McCarthy's protagonists do not use phones, listen to the radio or watch television, do not wash their clothes in washing machines, do not iron their shirts or vacuum their carpets. The most highly developed equipment they utilize are gas stoves, pistols, and pick-up trucks.

This total lack of advanced technology is obviously most conspicuous in McCarthy's latest novel, *The Road*, which tells an extremely poignant story of a father and son trudging through a barren country in a desperate search for food, warmth, and people who would view them as more than pieces of edible meat. In *The Road* everything is scorched, cold, dead, grey, and shapeless. From time to time a tremor shakes the desolate earth, which is covered with a newly fallen layer of dust each day. The bland sky, devoid of the sun, moon, and stars, is occasionally agitated by a sudden flash of lighting. We are not told what had brought about such a global cataclysm, but from the man's fragmentary reminiscences we may presume that it was a man-made disaster:

---

10 *MySpace*, "Cormac," http://profile.myspace.com/index.cfm?fuseaction=user .viewprofile&friendID=161785678 (accessed March 10, 2009).

> The clocks stopped at 1:17. A long shear of light and then a series of low concussions. He got up and went to the window. What is it? She said. He didn't answer. He went into the bathroom and threw the lightswitch but the power was already gone. A dull rose glow in the windowglass.[11]

The canister masks and biohazard suits that the bearded "bad guys" are wearing as well as the grayish snow and dark muddied water filling hollows in the ground may signify a nuclear strike—a catastrophe brought about precisely by means of man-developed technological advancement.

Technology in McCarthy's novels, present only through its conspicuous absence, is therefore certainly not an aid which makes human life simpler and easier, nor is it a sign of man's progress or upgrading. It is rather a major threat and a deadly enemy, not because it is alien and external, but precisely since it is our own creation. Like McCarthy's earlier novels, *The Road* also testifies to the human potential of self-destruction, the tendency to scapegoat and sacrifice weaker individuals, and the liability to succumb to our darkest drives and impulses. The telling difference is that the latest book uses the circumstances of a technological disaster, thereby highlighting a paradoxical clash with the increasingly "technologized" virtual reality of his readers, virtually communicating with and applauding McCarthy's virtual self.

Works Cited

Conlon, Michael. "Writer Cormac McCarthy Confides in Oprah Winfrey." *Reuters*, June 5, 2007. http://www.reuters.com/article/entertainment News/idUSN0526436120070605 (accessed March 10, 2009).

Garner, Dwight. "The Fighting Cormackians." *New York Times*, June 14, 2007. http://papercuts.blogs.nytimes.com/2007/06/14/the-fighting-cormackians/ (accessed March 10, 2009).

—. "Inside the List." *New York Times*, October 15, 2006. http://www.nytimes.com/2006/10/15/books/review/15tbr.html?_r=2&ref=review&oref=slogin&oref=slogin (accessed March 10, 2009).

Georgiades, William. "Required Reading." *New York Post*, June 3, 2007. http://www.nypost.com/seven/06032007/entertainment/required_reading_entertainment_william_georgiades.htm (accessed March 10, 2009).

McCarthy, Cormac. *The Road*. New York: Alfred A. Knopf, 2007.

New York Magazine. "Cormac McCarthy Bombs on 'The Oprah Winfrey Show.'" June 6, 2007. http://nymag.com/daily/entertainment/2007/06/cormac_mccarthy_slouches_towar.html (accessed March 10, 2009).

[11] McCarthy, *Road*, 45.

—. "Oprah's Cormac McCarthy Interview Is June 5." May 25, 2007. http://nymag.com/daily/entertainment/2007/05/oprah_to_interview_cormac.html (accessed March 10, 2009).

Raab, Scott. "The Coen Brothers' New Movie." Review of *No Country for Old Men*, directed by Ethan Coen and Joel Coen. *Esquire*, September 18, 2007. http://www.esquire.com/features/esquire-100/nocountry1007 (accessed March 10, 2009).

Woodward, Richard B. "Cormac McCarthy's Venomous Fiction." *New York Times*, April 19, 1992. http://www.nytimes.com/1992/04/19/magazine/cormac-mccarthy-s-venomous-fiction.html?sec=&spon= (accessed March 24, 2009).

How Far Do We Have to Go? Teleportation in the Service of American Cultural Communication in Octavia Butler's *Kindred* (1979)

Beata Zawadka

February 24, 2006 saw the death of the "grande dame of science fiction," Octavia Butler. A native of Pasadena, California, born on June 22, 1947, this only child of a black shoeshiner and a white people's maid became a well-known science fiction writer with the publication of the first three novels of her five-volume "Patternist" series in the late 1970s.[1] Since then, she went on to write a dozen or so novels and numerous short stories, and won two Hugos and a Nebula. Her *Parable of the Sower*, the first of her "Parable" series, published in 1993, was also shortlisted for the New York Times Notable Book of the Year award. In 1995, one of a handful of widely-known black science fiction writers and the only black woman to have made her living writing science fiction full time, was awarded the prestigious MacArthur Foundation fellowship, commonly referred to as the "genius grant." This award, given to a science fiction writer for the first time, ultimately confirms her reputation as an inspiration not only to science fiction but to all kinds of American writers and audiences.

The book that, unlike the rest of her artistic legacy mentioned above, did not so much contribute to but actually made this reputation is Butler's 1979 novel entitled *Kindred.* It tells the story of a young middle-class African American heroine, Dana, who is repeatedly transported between 1976 Los Angeles and antebellum Maryland, where her slaveowning forebear's plantation is situated. Over the course of the novel we learn that these mysterious transportations across the centuries are to ensure Dana's

[1] The entire Patternist series is composed of the following novels: *Patternmaster* (1976), *Mind of My Mind* (1977), *Survivor* (1978), *Wild Seed* (1980), and *Clay's Ark* (1984). It was compiled in 2007 as *Seed to Harvest*.

own birth as a descendant of a white man and a black woman as well as her present existence as both a happy wife to Kevin (who is white) and a very promising writer. For freedom and love then, not only is Dana to consciously accept being enslaved in the past whenever the life of Rufus Weylin, her white ancestor, is endangered (which, back in the South of the early 1800s happens frequently). She must also—and equally consciously—arrange for the rape of the free black woman by the white man, thereby arranging, too, for her enslavement and consequently coming face to face with her own slave heritage. The questions of slavery versus freedom, past versus present, racial inequity, politics, sexuality, or sexual identity, all intertwined in *Kindred* through the concept of time travel, are thus the main reason for which Octavia Butler has been considered a national inspiration; in this book she demonstrated herself as capable of both communicating and making smoothly communicable various, apparently mutually exclusive cultural messages long before cultural studies opted for such a strategy to be their own, officially recognized methodology.

This paper, very much a homage to the great writer on the forthcoming first anniversary of her death, aims at presenting *Kindred* exactly as Butler's comment on the future, the 21st century nature of the American cultural practice. In my opinion, she sees this practice as a radically integrative strategy that, in its refusal to separate what seems to be incompatible values, and in its attempt not to include them in one unified picture,[2] resembles, both formally and in subject matter, the contemporary telecommunication technologies such as the Internet and television, or Web communication practice. By this I mean that, like these technologies, nowadays considered the most democratic means of cultural communication, *Kindred* is capable of displaying antithetical values to enable the processing of possible traumas which result from learned convictions that the world can be described by dichotomies.[3] This is possible because *Kindred*, like the telecommunication technologies in question, allows its readers to move beyond their individual alienations and delight instead in feeling an integral part of the "global" family, which, in turn, points to future America's capacity to imagine empathy and meaningful intercultural unions or else, to feel interculturally "kindred."

---

[2] Tara McPherson refers to such practices collectively as performing "lenticular logic" and she differentiates between its separatist and additive forms. See McPherson, *Reconstructing Dixie*, 24–28, 98; cf. ibid., 86–92.

[3] See Wexelblat, "Auteur in the Age of the Internet," 209.

The novel already suggests as much on the formal level. I refer here to the fact that, apart from being pigeonholed first as a classic science fiction masterpiece and, later, with the rise of African-American studies in the 1980s, as a "distinctive contribution to the genre of neo-slave narrative,"[4] *Kindred*—in the eyes of its author—remained till the end of her life a "kind of grim fantasy" and hence, a narrative creation in its own right.[5] Butler's obstinacy in formally defining *Kindred* seems justified. For one thing, it is difficult to classify the novel solely as a science fiction piece because, although it indeed employs the concept of time travel as a method of storytelling, it does not present this method as a well-engineered and well-motivated mechanism, so typical of Wells's stories. Dana, the protagonist of *Kindred*, describes her time traveling as simply starting to feel "dizzy, nauseated," losing sight of the surrounding world: "[t]he room seemed to blur and darken around me" and, finally, losing consciousness: "[t]he house, the books, everything vanished."[6] According to her sensibility then, the experience resembles fainting and as such it has more to do with seemingly mechanism-free and hence mysterious, irresistible forces operating in the mind such as teleportation that simply implicates people in itself than with externally controlled processes. This, in turn, renders the narrative mode of *Kindred* less science-fiction-like and closer to that employed by the fantasy genre.[7]

What is more, Butler uses these forces to move Dana not so much in the direction of the future so that she can verify the truth of her journey as something with a political and sociological edge that is characteristic of not only science fiction but also of slave narrative, another of the novel's formal constituents which it refuses to follow blindly. Rather, judging by the fact that Dana journeys from the contemporary American Southwest to the antebellum South and therefore, from the freedom-oriented period and region into the slave past and space—literally in the direction opposite to that typically followed by the protagonist of a slave narrative—Butler makes use of fantasy to invert such a narrative and hence, to metaphorically enslave it so as to free it anew. Accordingly, although *Kindred*, as an inverted slave narrative still remains propagandist literature, in the manner of its "straight" predecessors—as exemplified in

---

[4] See Crossley, "Critical Essay," 265.

[5] Butler, "Interview: Octavia E. Butler," an interview by John C. Snider.

[6] Butler, *Kindred*, 13. All the subsequent quotations from *Kindred* in this paper are from this edition of the book.

[7] For the differences between the two genres see Flynn, "Time Travel Literature"; cf. Frelik, "Proza popularna," 111–12.

the fact that with a young black female writer as the protagonist[8] it appeals to both feminist and ethnic circles—it simultaneously declares all formal "straightness" a fantastic given and thus problematizes its further production as the only road to truth.

As such, the narrative mode that Butler creates out of the two already existing ways of storytelling demonstrates a radically integrative cultural strategy similar to contemporary telecommunication technologies not only because, through it, Butler processes the opposing cultural currents (in this case: the popular versus the academic) so that they become a brand new quality. This is also because, in so doing, Butler additionally allows for this quality to engage in conversation with various other, conservative and progressive elements in American culture, thereby building a whole web out of their geographies, which affords different possibilities for autonomy or resistance. One good example of this is the way in which the form of the book meets and deals with its own content. Namely, as a "grim fantasy" and so, by definition, half a tale about magical powers and half a horror story,[9] *Kindred* stages an understanding of its own content that, by positioning its characters amid what is both "kin" and "dread" to them so that they themselves and these notions cannot but become the web of the "familiar," simultaneously allows a perception of this positively marked structure to be also perceived in terms of boundedness, thus addressing the meaning of boundary and transgression. In this sense, at its deepest level *Kindred* is primarily a book about the need to constantly redefine familiarity anew, or rather to explore it—like, for example, the Web users do—as the web of both affection and power, for only then can these two elements fundamental to all effective human relations begin struggling towards new connections other than themselves.

It is for this reason, I believe, that the underlying familiar opposition which the novel tackles anew is that of slavery versus freedom. On the one hand, judging by the many scenes in the book in which slaves on the Weylin plantation are treated brutally by the whites and scenes in which they attempt to oppose this brutality by escaping North, finding refuge amid the community of slaves, or in white female domestic space, *Kindred* presents the bondage-freedom opposition exactly in the way which the contemporary reader, viewer or Web searcher expects. Namely, it is both a true "horror story" and one of truly magical hopes and possibilities that implicated only Afro-Americans and hence, a web of very clearly defined

---

[8] On the characteristics of the traditional slave narrative, see e.g. Porter, "Social Discourse," 345–63.

[9] *SciFipedia*, http://scifipedia.scifi.com/index.php/Dark_fantasy (accessed October 28, 2007; site now discontinued).

power and affection. On the other hand, however, *Kindred* also implies bondage as too complex a phenomenon to be uniformly "fixed" and as such, "free" to displace the "true" comforts of its more familiar versions. This is demonstrated by the way Dana—a captive in a process that she cannot control and that additionally enslaves her in the antebellum South and on a slave plantation—perceives this enslavement when she first returns to the present and tries to explain her disappearance to her husband: "It's becoming like something I saw on television or read about—like something I got second hand."[10] In this way, she suggests that Afro-Americans have perceived the experience of bondage as more "virtual" than "real," and therefore as a web of not so clear-cut divisions at all.

To my mind, the novel shows this most convincingly via exploring its central pairing, that is, the tight bond between Dana and her white forefather, Rufus. On the surface, this bond, since it is unconsciously produced and controlled by the white man—for Rufus conjures Dana up as a guarding angel of sorts when he is in need so that she is forced to mammy him in order to save her own life and that of her family—looks like a typical old southern white oppressor/black victim relationship, inevitably connected with power. This impression is consciously confirmed on Dana's part, too, as throughout the book she is struggling against Rufus's control over her, be it by befriending him so that he becomes a kinder master than his father, or even by cutting her wrists at one point to free herself from him. Yet, as Rufus grows into adulthood, it becomes clear that no kind of sculpting can remake his relationship with Dana, for, as time goes by, he simply demands more and more of it. Dana also finds that, apart from rebelling against him, she wants to give him comfort, no matter how repulsed she is by Rufus's distinctly old southern white master-like actions; this capacity to care for someone whom she perceives as primarily her tormentor surprises her: "I always seem to forgive him for what he does to me. I can't hate him the way I should . . . ."[11] The fact that Dana, in trying to name her feelings towards Rufus—"What we had was something new, something that didn't even have a name"—ascribes them to "some matching strangeness"[12] in both him and herself, thus provides us with a definition of slavery as interdependency of power and affection rather than sheer tyranny of one race over another. Therefore, clearly, bondage is indicated as an access route not only to black but also white pain.

---

[10] Butler, *Kindred*, 17.
[11] Butler, *Kindred*, 224.
[12] Ibid., 29.

As such, *Kindred* refuses to be interpreted, like many slave narratives, as a mere power struggle story—although it inevitably gravitates towards one—and begins to portray much messier terrains of the old southern slaveholding society, that is its psychosexual dimensions, concerning affection and intimacy. I refer here primarily to the fact that even though the novel demarcates the lines of power in the Weylin plantation household—which are relatively predictable: black versus white, male versus female, or free versus enslaved[13]—it nevertheless does not sidestep the problem of interracial emotional and physical relations such as, for example, Rufus's love for black women that other popular culture works of the *Gone With the Wind* type carefully omit. Neither does *Kindred* protect its readers against the knowledge about consequences of such relationships, which are always destructive for both parties. The fact that Dana is able to sever her ties to Rufus and free herself from the 1800s only by killing him during an attempted rape, and that she loses her own arm in the process (the part which Rufus grasps, from the elbow down, remains stuck in the plaster of her living room wall, scarring her for life) may not only imply that slavery as an interracial social phenomenon is about emotional and physical disfigurations but that the actual range of these disfigurations transgresses time and space long after slavery itself ends.

This, in turn, is suggested in the book by the relationship of Dana and her white husband Kevin, a tie that actually mirrors the relationship of Dana and Rufus, precisely because, according to Marc Steinberg, it represents, as a marriage, a "classic type of human ownership in western civilization."[14] Hence, this bond implicates the characters in the same gender/class/race-oriented, supremacist culture as the one governing the world which the planters built. *Kindred* symbolizes this, for example, through its presentation of the physical similarity, language and accent shared by Kevin Franklin and Rufus Weylin to the degree that after a time they merge in Dana's mind so that she mistakes one for the other: "The words echoed strangely in my head. Kevin had said something like that to me once. I opened my eyes again to be sure it was Rufus."[15] Thereby Dana also confuses the relationship with her husband with the one she has with her historical oppressor, and consequently, points to the lingering synonymy of the meanings of the words "husband" and "master." Even though Kevin's behavior seemingly posits him far from the mastery performed in the grand old southern style—for he is also described as a

---

[13] On the exceptions from this rule concerning, for example, the role and position of the mistress in the Weylin household consult McPherson, *Reconstructing Dixie*, 89.

[14] Steinberg, "Inverting History," 469.

[15] Butler, *Kindred*, 213–14.

liberated, modern man who loves his wife, loathes the chattel system and works with the Underground Railroad when trapped in the past—he also admits that he could get used to these times. However, this dissimilarity to Dana's white forefather can be interpreted to the effect that he is simply as much a man of his time as Rufus is of his own, and hence unable to transform the painful gender and race link with the past that has ideologically marked the white elite man as America's chief oppressor.[16]

Thus, on the one hand, via the marriage metaphor, *Kindred* traces the mentally deforming pressures of slavery as a masterful institution, that is one that continues to deny its status as a culturally shared ground. The novel exemplifies this further by showing that even in late twentieth-century America and in a large, progressive and cosmopolitan city such as Los Angeles, Kevin and Dana's interracial marriage is still not fully sanctioned: the two must endure such comments as "the weirdest-looking couple,"[17] or even "chocolate and vanilla porn"[18] on the part of their co-workers. What is more, such racism penetrates far deeper than the level of mere community relations, for Dana and Kevin's marriage is primarily rejected on the level of the family, thereby confirming this institution as still crucial in reproducing the stuff slavery is made of, namely hatred and power and, in consequence, also dulled, slave-like contemporary sensibilities. In effect, it leaves people no choice for encounters other than work relations, as is the case with Dana and Kevin, who first meet when employed by a blue-collar temp agency, commonly referred to by its workers as the "slave market" and whose laborers Dana depicts as "nonpeople rented for a few hours."[19] Thus, it is suggested that, like slavery in the past, the telecommunication technologies that Butler's book represents figuratively—and hence, contemporary American culture—deforms its culturally differentiated participants by making them visible to one another only in terms of the economy.

On the other hand, such a reading of *Kindred* not only clearly bespeaks the lingering refusal to recognize the uniqueness of the American identity as a multicultural being but it also provides ways to understand why this situation persists. This is because, in interpreting the novel as a critique of

[16] In the book, this enduring pattern of the white elite male's dominance and his inability to transcend it might also be suggested by the fact that, when traveling with Dana back to the past, Kevin remains trapped there much longer than his wife and when he is finally released, it also takes him longer to accommodate himself back in the present.

[17] Butler, *Kindred*, 57.

[18] Ibid., 55.

[19] Ibid., 53.

American cultural practices which fail due to the idiosyncrasies of the American economic system, the reader begins to doubt whether the typical "literacy—identity—freedom" paradigm that American culture propounds as the only way out of economic misery and towards making sense of the world can actually be liberating at all.[20] This paradigm to which Dana resorts as a way of making the most of her situation when she realizes she will continue to travel through time until she ensures that Rufus grows up to initiate her family line, and which she decides to use for the sake of teaching the slave children around her to read and write—and to run for freedom as soon as they can—actually has quite opposite consequences. Namely, even though this paradigm gives a slave some freedom—as it happens with Dana who, when caught with her pupils and books, is so mercilessly beaten by Rufus's father that she falls unconscious and feels pulled back to California—it nevertheless fails to secure an individual identity with no connection whatsoever to the society that enslaved the slave. After all, Dana lands in the present with deep wounds on her back that take a long time to heal into scars, thereby remaining a visible mark of representing the subjectivity whose contours are not chosen by an individual but rather delineated by abstract forces. To this we should add the fact that, on returning to antebellum Maryland several weeks after the beating, Dana learns that her disappearance prompted old Weylin to punish her fellow slaves by selling some of their family members away from the plantation, thus making subjectivity a communal pain. Clearly, no matter how carefully exercised, the paradigm in question will always undo its followers and prove successful in preventing American multiculturalism from flourishing as long as it is believed to be transcending human relations with the world into something "real" or "true" rather than a literally and metaphorically "escapist" *modus operandi* indispensable for dealing with this world.

Although *Kindred* questions education as a way to a healthy identity and freedom, this does not mean that, simultaneously, via the conceit of time travel, it does not treat education as a springboard to understanding slavery—and by extension contemporary American cultural practice—as a process of constant accommodation to change and hence a route to reawakening imagination and to liberating America's creative potential. I am referring here to the scene in which Dana, having resorted to book reading upon her return to California after the beating mentioned above, starts to reassess her own relation to slavery as no longer a strange

[20] For the role of this paradigm in manipulating history, see Yaszek. "'A Grim Fantasy,'" 1–10.

personal experience that alienates her from the rest of the world and thus "fixes" her in the position of this world's "other," but a part and parcel of a broader—American and wider—experience as a whole and therefore its mobile element, capable of fitting diversified "gaps":

> I read books about slavery, fiction and nonfiction. I read everything I had in the house that was even distantly related to the subject—even *Gone With the Wind*, or part of it. But its version of happy darkies in tender loving bondage was more than I could stand. Then, somehow, I got caught up in one of Kevin's World War II books—a book of excerpts from the recollection of concentration camp survivors. Stories of beating, starvation, filth, disease, torture, every possible degradation. As though the Germans had been trying to do in only a few years what the Americans had worked at for nearly two hundred. The books depressed me, scared me, made me stuff Kevin's sleeping pills into my bag. Like the Nazis, ante bellum whites had known quite a bit about torture—quite a bit more than I ever wanted to learn.[21]

The "mobility" that Dana's newly gained knowledge gives her seems unbearable to her at first, forcing her to face what she should have already acknowledged had she taken a closer look at the direction of her time traveling, namely that progress is never a certainty. However, exactly by allowing her to see immobility as also the *status quo* and hence, the "real" thing, it also helps Dana, albeit unconsciously, forge education into a connection with the past as well, and so, into a refusal to accept this past as merely a comfortably distanced science fiction story or a slave narrative.

In this light, the "grim fantasy" genre that *Kindred* represents as an integrative strategy similar formally and in its subject matter to modern telecommunication technologies, and which is a compromise between the narrative strategies of both a time travel story and a memoir of a former slave, can also be read as a suggestion on how to approach "realities"—historical and other—so that they no longer disturb the perception of American cultural practice as it has always been, namely, as a multidimensional phenomenon. By this I mean that, much as the novel links past and present so that the two collide in the narrative becoming one, it also allows its protagonist to retain them as her two separate—although transformed—histories and thus let both custom and imagination speak through each of them. This is most clearly suggested in the epilogue to *Kindred* in which Dana, after the arm that she lost on her final "trip" home has sufficiently healed, decides to fly to present-day Maryland so as

[21] Butler, *Kindred*, 116–17.

to search for the remnants of the Weylin plantation there. Once in Maryland, she learns from an old newspaper article that the plantation has been destroyed, which pushes her towards the Maryland Historical Society to find out about her subsequent family history, only to realize that no official records of it exist. When Kevin, who accompanies her on this trip, suggests that she will "probably never know" exactly what happened to her ancestors, Dana "touche[s] the scar Tom Weylin's boot had left on [her] face, touche[s] [her] empty arm sleeve and replies 'I know.'"[22] In this way, by affirming what is nonexistent as if it were real, Dana not only acknowledges history as a part of the present, thus commencing a new tradition, but she also defines this tradition as primarily based on personal history and hence on the creative inspiration necessary for producing what, in terms of culture, connects people most—stories.

At the end of the book Dana surfaces as the harbinger of a new mode of American cultural practice, one that 1970s readers may have indeed regarded as something out of a fantastic science fiction scenario, i.e. the modern telecommunication technologies of the Internet variety. This is not only because, her full name being Edana, she struggles throughout the novel to be called Dana instead; the protagonist of *Kindred* resembles an electronic impulse-like construct as much as a typical book character. With the novel beginning where it actually ends—that is, at the moment Dana loses her arm, a moment which Octavia Butler never attempts to rationalize in the book—the readers who decide to follow Dana cannot but feel that they are implicated in some sort of a mystery, a twilight zone between reality and unreality that is closer to virtual than actual time and space. Dana's venture begins, in real time, in California on June 9, 1976, the day of her twenty-sixth birthday and ends on July 4, 1976, the Declaration of Independence bicentennial, a period that equals several years spent by Dana in the Maryland of fictional time. As such, it provides an insightful comment—much as contemporary telecommunication technologies do—on the actual limits of time and space, and thus on the limits of our identity. Therefore, it then becomes clear that in making her protagonist "E-Dana," Butler pushed us all, too, via *Kindred*, towards processing the American cultural practice—and ourselves—into representing "a kind of grim fantasy" of a digitalized, global community.[23]

---

[22] Ibid., 264.

[23] Crossley, "Critical Essay," 273. A possible confirmation of the above is the fact that in 2004, twenty five years after its first publication, *Kindred* was reedited to celebrate its emergence as the leader in the 2003 local competition titled "If All of Rochester, New York Read the Same Book." This new edition immediately sold millions of copies.

Works Cited

Butler, Octavia E. "Interview: Octavia E. Butler." By John C. Snider. *SciFiDimensions*, 2004. http://www.scifidimensions.com/Jun04/octaviae butler.htm (accessed March 10, 2009).

—. *Kindred*. Boston: Beacon Press, 2004.

Crossley, Robert. "Critical Essay." In *Kindred* by Octavia E. Butler. Boston: Beacon Press, 2004.

Elliot, Emory, ed. *Columbia History of the United States*. New York: Columbia University Press, 1988.

Flynn, John L. "Time Travel Literature." http://pages.towson.edu/ (accessed October 16, 2007).

Frelik, Paweł. "Proza popularna—science fiction." In *Historia literatury amerykańskiej XX wieku*, edited by Agnieszka Salska, 2 vols. Kraków: Universitas, 2003.

McPherson, Tara. *Reconstructing Dixie: Race, Gender, and Nostalgia in the Imagined South.* Durham and London: Duke University Press, 2003.

Porter, Carolyn. "Social Discourse and Nonfictional Prose." In *Columbia History of the United States*, edited by Emory Elliot, 345–63. New York: Columbia University Press, 1988.

Steinberg, Marc. "Inverting History in Octavia Butler's Postmodern Slave Narrative." *African American Review* 38, no. 3 (Autumn 2004): 467–76.

Wexelblat, Axel. "An Auteur in the Age of the Internet: JMS, *Babylon 5*, and the Net." In *Hop on Pop: Politics and Pleasures of Popular Culture*, edited by Henry Jenkins, Tara McPherson, and Jane Shattuc, 209–26. Durham and London: Duke University Press, 2002.

Yaszek, Lisa. "'A Grim Fantasy': Remaking American History in Octavia Butler's Kindred." *Signs: Journal of Women in Culture and Society* 28, no. 4 (2003). http://www.journals.uchicago.edu (accessed October 18, 2007).

THE FUTURE GROWLS: VOICES AND MACHINES IN EUDORA WELTY'S *ONE WRITER'S BEGINNINGS*

NOEL POLK

Eudora Welty begins her 1983 memoir, *One Writer's Beginnings,* with an untitled one-paragraph prelude; she separates it from the rest of the book's three parts by placing it before the title for Part I, "Listening," in effect asking us to take it as expressing something quintessential about her relationship with her parents and, perhaps, about the structure of all families, a subject central to her work:

> *When I was young enough to still spend a long time buttoning my shoes in the morning, I'd listen toward the hall: Daddy upstairs was shaving in the bathroom and Mother downstairs was frying the bacon. They would begin whistling back and forth to each other up and down the stairwell. My father would whistle the phrase, my mother would try to whistle, then hum hers back. It was their duet. I drew my buttonhook in and out and listened to it—I knew it was "The Merry Widow." The difference was, their song almost floated with laughter: how different from the record, which growled from the beginning, as if the Victrola were only slowly being wound up. They kept it running between them, up and down the stairs where I was now just about ready to run clattering down and show them my shoes.*[1]

In this elegant little fable, Daddy is upstairs doing what men are supposed to do, getting ready for work, while mother is downstairs doing what women are supposed to do, preparing breakfast for him. The child Eudora positions herself on the stairs between them, doing what children do, as she would have it: spying. Buttoning her shoes is a ruse. The curious phrase, "*When I was young enough to still spend a long time buttoning my shoes in the morning,*" suggests that she thinks that her position involves some risk. If her parents discover her there, they might, correctly, assume that she is spying on them, listening for something they don't want her to

[1] Welty, *One Writer's Beginnings*, 837.

know. Her parents, of course, do not seem to be aware that she is spying. The episode is thus about how the child Eudora experiences herself in relation to her parents or, rather, about how the older Welty remembers that relation—a relationship she explores throughout her work.

This episode is based in the child Eudora's assumptions that her parents' communication, whatever it is, has nothing to do with her, that they are communicating some secret that she is no part of and that it is vitally important to figure out what they are saying to each other and not to her. When she cannot figure it out, she disrupts her way into their communication with her own clattering. Her real career as a fiction writer begins when she finally understands that she's much more likely to hear and understand if she just listens, maintaining her presumptive—even if phony—neutrality in the triangle, like on the stairs buttoning her shoes. She's much more interested in what she doesn't hear, in what she *might* hear if she just listens hard enough, than in what she does hear. Often what she does hear as a child is the murmur and drone of voices barely audible. When she would finally convince them that she was asleep, her parents would drape

> the lampshade with a sheet of the daily paper, which was tilted, like a hatbrim, so that they could sit in their rockers in a lighted part of the room and I could supposedly go to sleep in the protected dark of the bed. They sat talking. What was thus dramatically made a present of to me was the secure sense of the hidden observer. . . . What they talked about I have no idea, and the subject was not what mattered to me. It was no doubt whatever a young married couple spending their first time privately in each other's company in the long, probably harried day would talk about. It was the murmur of their voices, the back-and-forth, the unnoticed stretching away of time between my bedtime and theirs, that made me bask there at my distance. What I felt was not that I was excluded from them but that I was included, in—and because of—what I could hear of their voices and what I could see of their faces in the cone of yellow light under the brown-scorched shade. . . . I have always been shy physically. This in part tended to keep me from rushing into things, including relationships, headlong. Not rushing headlong, though I may have wanted to, but beginning to write stories about people, I drew near slowly; noting and guessing, apprehending, hoping, drawing my eventual conclusions out of my own heart . . ."[2]

In the listening, then, she places herself in the position of what she calls a "privileged observer," a position which allowed her to get distance, "a

[2] Ibid., 861–63.

prerequisite of [her] understanding of human events."[3] She's always on those stairs as a fiction writer, buttoning her shoes, trying to figure out what people are saying to each other, how they say it, what mysterious worlds those private communicative sounds create in the air around her head, waking or sleeping—at home, in the car, on the bus or train. What she hears and doesn't hear—what the words say and don't say—become a kind of yeast for some of America's finest fiction.

On the stairs she's listening to her parents' happy if clumsy duet, a clumsiness born of differences easily overcome, at this early point in their marriage, by their good humor and their love; her mother can't return her father's melody with a whistle, so she returns it by humming. Whistling and humming would seem to be incompatible harmonically, but their duet nevertheless "float[s] with laughter" as they prepare for the day. They are humming a well-known song from Franz Lehar's operetta *The Merry Widow*, a song which Welty, writing several decades afterwards and for her own literary and personal reasons perhaps imposes upon this scene. *The Merry Widow* loads *One Writer's Beginnings* with irony: barely two decades later, Welty's father died, leaving her mother *anything but* a merry widow; how un-merry she is for the rest of her life becomes the subject of much of *One Writer's Beginnings* and of her novel *The Optimist's Daughter*. What appears here as a *frisson* of their differences, then, becomes, as we read, a friction that their love cannot, in the long term, overcome. Their laughter, their pleasure in each other, so palpable here, gets complicated over time in the very act of living.

The episode ends, in fact, as Eudora forces herself between them: "*I was now just about ready to run clattering down and show them my shoes*." The "*clattering*" noise of her shoes disrupts the floating laughter of their music; to her older self, looking back, her presence alters the young lovers' harmonies and their joy, though she does not depict them here as being in any way discomfited by or amused at or even aware of her listening or her interruption. The episode is thus more nearly about her sense of herself as an intruder than about what's going on between her parents. She makes herself a disruption of their love, a part of their love that changes it necessarily. Though she is a direct result of their love and a happy part of it, their love can never be the same as before she was born. That she depicts herself at this age as an outsider to their love, to their intimacies, may be a crystallization of her lifelong commitment to her artistic position as a privileged, but outside, observer.

---

[3] Ibid.

Though we do not actually hear from the Victrola in this passage, Welty makes of it here and later a troubling fourth member of her family. Among the earliest devices to reproduce the human voice, indeed to mass-produce it commercially, the Weltys' Victrola must have been among the earliest ones since they were produced between 1901 and 1929; their owning one reflects her father's interest in gadgets and the future. In a passage soon to follow, Welty notes that her mother "always sang to her children." "Her voice came out just a little bit in the minor key. 'Wee Willie Winkie's' song was wonderfully sad when she sang the lullabies." Her father, responding to the sadness and no doubt wanting to protect his daughter even from that, replaces his wife's voice with the Victrola's mechanical reproduction of a professional singer's voice, replaces the mother's real voice with a mechanical one that begins with a growl: "Oh, but now there's a record," Welty imagines her father saying. "'She could have her own record to listen to,' my father *would have said*." He apparently did *not* actually *say* that, but might as well have. Her father thus intervenes between Eudora and her mother: "all of Mother's lullabies . . . could be played to *take her place* [italics mine]. Soon I was able to play her my own lullabies all day long." We are left to wonder how often he intervened in such situations and how the intervention affected her mother—to be told that your voice is not good enough to sing lullabies to your own daughter? Does she switch from whistling to humming because she fears her husband will find even her whistling to be not good enough? How many such minor indignities does it take to undermine love? How do they chip away at joy?

Whatever the effect on her mother, Welty records her own joy in the technology that takes her mother's place:

> Our Victrola stood in the diningroom. I was allowed to climb onto the seat of a diningroom chair to wind it, start the record turning, and set the needle playing. In a second I'd jumped to the floor, to spin or march around the table as the music called for—now there were all the other records I could play too. I skinned back onto the chair just in time to lift the needle at the end, stop the record and turn it over, then change the needle. That brass receptacle with a hole in the lid gave off a metallic smell like human sweat, from all the hot needles that were fed it. Winding up, dancing, being cocked to start and stop the record, was of course all in one the act of *listening*. . . . Movement must be at the very heart of listening.[4]

[4] Welty, *One Writer's Beginning*, 848–51.

Christian Welty

> loved all instruments that would instruct and fascinate. His place to keep things was the drawer in the "library table" where lying on top of his folded maps was a telescope with brass extensions, to find the moon and the Big Dipper after supper in our front yard, and to keep appointments with eclipses. There was a folding Kodak that was brought out for Christmas, birthdays, and trips. In the back of the drawer you could find a magnifying glass, a kaleidoscope, and a gyroscope kept in a black buckram box, which he would set dancing for us on a string pulled tight. He had also supplied himself with an assortment of puzzles composed of metal rings and intersecting links and keys chained together, impossible for the rest of us, however patiently shown, to take apart; he had an almost childlike love of the ingenious.[5]

He gives his sons an electric train set for Christmas: "All of this, . . . especially the train, represent[ed her] father's fondest beliefs—in progress, in the future. With these gifts, he was preparing his children."[6]

"And so was my mother," Welty insists, in a separate, one paragraph sentence for emphasis, "with her different gifts." Her gift to her children was in essence backlooking—another of their curious duets. She gave them her love of books, and a houseful of them: encyclopedias, dictionaries, "the Unabridged Webster, the Columbia Encyclopedia, Compton's Pictured Encyclopedia, the Lincoln Library of Information, and . . . the Book of Knowledge," and the 25th edition of the Britannica, which her father, "his face always deliberately turned toward the future, was of course disposed to think better than any previous edition."[7] His view to the future was a manic and perhaps slightly ridiculous need to protect them all against eventuality—he carried a ropeladder with them on their automotive trips to have in case there was a fire in the hotel, for example!—a need to control eventuality that seems to have spilled over into a need to control their daily lives, even their pleasure in singing. But he could not anticipate or prevent the illness, or the treatment, that took his life in 1931, at the age of 52.

By 1931 the science of the blood bank had not yet made it to the United States and, indeed, the science of blood transfusion itself was in its infancy, in spite of a history that dated back some centuries in Europe and in spite of advances that had been made during World War I. Only the faintest rumors of blood types, blood mixing, and methods of transfusion

---

[5] Ibid., 839.
[6] Ibid., 840–41.
[7] Ibid., 842.

had reached Mississippi by 1931, when her father fell ill of a disease that neither Welty nor her biographer, Suzanne Marrs, identifies,[8] but which Ann Waldron,[9] without attribution, says is leukemia. With no knowledge of the need to match blood types, the doctors proposed a simple straight transfusion from Welty's mother, who insisted that she be the donor of the blood. As Welty watched, her mother lay down parallel to her father on a separate cot: "Then a tube was simply run from her arm to his. . . . All at once his face turned dusky red all over. The doctor made a disparaging sound with his lips, the kind a woman knitting makes when she drops a stitch,"[10] as though her own mother might have thought that saving her husband's life was as easy as knitting. What she thought was life moving from her arm to his was actually a co-conspirator with the death that was already so near.

Part of her mother's need to save his life was that he had saved hers, earlier, when after the birth of one of their children she contracted septicemia, then "nearly always fatal."[11] Her father conceived a remedy, a "measure . . . out of the strength of desperation," which told him that champagne was a magic elixir that would save her. But of course champagne was not sold in Jackson in those days. Undaunted, he used the telephone, also relatively new in Jackson, to call an Italian orchard grower in Canton, a rural village about 30 miles north of Jackson. He naturally had the train schedule handy and told Mr. Trolio which train to put it on; he met the train at the station: Mr. Trolio "sent the bottle in a bucket of ice and my father snatched it off the baggage car. He offered my mother a glass of chilled champagne and she drank it and kept it down. She was to live, after all."[12] They were lucky, of course: her survival was completely coincidental with the champagne, which could have had no effect whatsoever on her septicemia. But the phone and the precious train schedules allowed them the illusion that his faith in his own capacity to stave off the unexpected and in communication technology had played a part in saving her.

In this instance, the telephone did its illusionary work. In *One Writer's Beginnings*, it works as a communicative arm of the redoubtable Miss Duling, the fiercely dominating teacher and principal of Davis School: for her the phone served the same peremptory service as the loud daunting

---

[8] Marrs, *Eudora Welty: A Biography*, 34.
[9] Waldron, *Eudora*, 47.
[10] Welty, *One Writer's Beginning*, 935–36. See Prenshaw's elegant discussion of this scene in "Construction of Confluence," 186–87.
[11] Welty, *One Writer's Beginning*, 857.
[12] Ibid.

bell that she rang to call the children to school and to send them home, to dismiss them for recess and lunch and then bring them back to the classroom. Miss Duling, like her father "a lifelong subscriber to perfection,"[13] used the telephone as an instrument of her indomitable will, as did her father in ordering the champagne. The phone was for her an extension of the bell at the end of her arm that she used to call to attention all the state and city leaders she had taught during her three-generation rule at Davis school, just as if they were still her students:

> When she wanted something done—some civic oversight corrected, some injustice made right overnight, or even a tree spared that the fool telephone people were about to cut down—she telephoned the mayor, or the chief of police, or the president of the power company, or the head doctor at the hospital, or the judge in charge of a case, or whoever, and calling them by their first names, *told* them. It is impossible to imagine her meeting with anything less than compliance.[14]

One other telephone appears in *One Writer's Beginnings* that has quite a different effect on the young Eudora—a typical incident, one takes for granted—when her mother received a call from one of her friends

> who always talked a long time. I knew who it was when my mother would only reply, now and then "Well, I declare," or "You don't say so" or "Surely not." She'd be standing on the wall telephone, listening against her will, and I'd sit on the stairs close by her. Our telephone had a little bar set into the handle which had to be pressed and held down to keep the connection open, and when her friend had said goodbye, my mother needed me to prize her fingers loose from the little bar; her grip had become paralyzed. "What did she say?" I asked.
>
> "She wasn't *saying* a thing in this world," sighed my mother. "She was just ready to talk, that's all."[15]

She's thus listening again, spying—though this time listening to only one side of the conversation. But she's much less interested in her mother's bored minimal responses than in what the caller on the other end of the phone is saying, whose side of the conversation she can't hear and can only imagine: "Years later," she writes, "beginning with my story 'Why I Live at the P.O.,' I wrote reasonably often in the form of a monologue that

---

[13] Ibid., 863.
[14] Ibid., 863–64.
[15] Ibid., 852–53.

takes possession of the speaker. How much more gets told besides!"[16] Nobody would confuse a monologue with communication.

Given her father's absorption in gadgets and technology, it is perhaps to be wondered that Welty didn't share her father's sense of the modern. Very little of technology appears in her fiction and when it does it usually does so with a growl of one sort or another that problematizes it. In *The Ponder Heart*, Uncle Daniel telephones Edna Earle with fear and trembling in his voice to order her to come out to the Ponder plantation to save him from his wife. In "Why I Live at the P.O." Edna Earle, who has been in constant miscommunication with her family, is finally driven to take her belongings, move out of the family home, and move to any community's communication center, the post office; she drags with her the radio and listens to the war news. There's a radio in a very early story, "Magic," never collected, which plays a sickening Hawaiian music, the "grovelling sweetness of the guitars"[17] that evoke the faraway romantic islands while a family sits "under a fringed lamp, their heads sunk as though all their necks had been broken."[18] Likewise in "Old Mr. Marblehall" the citizens of Natchez listen to a radio in the evening that "seems to mourn and cry for them."[19] Perhaps the job Welty took publishing the newsletter for Jackson's station WJDX spoiled the radio for her! It's also worth noting, finally, that as a girl she took some pleasure in the dictaphone in her father's office.

There's not much of the movies in her work either, but in *One Writer's Beginnings*, she uses a movie to explain another of her mother's legacies.

> When my mother would tell me that she wanted me to have something because she as a child had never had it, I wanted, or I partly wanted, to give it back. All my life I continued to feel that bliss for me would have to imply my mother's deprivation or sacrifice. I don't think it would have occurred to her what a double emotion I felt, and indeed I know that it was being unfair to her, for what she said was simply the truth.
>
> "I'm going to let you go to the Century Theater with your father tonight on my ticket. I'd rather you saw *Blossom Time* than go myself."
>
> In the Century first-row balcony, where their seats always were, I'd be sitting beside my father at this hour beyond my bedtime carried totally away by the performance, and then suddenly the thought of my mother staying home with my sleeping younger brothers, missing the spectacle at this moment before my eyes, and doing without all the excitement and

---

[16] Ibid., 853. Welty, "Why I Live at the P.O.," 69.
[17] Welty, "Magic," 6.
[18] Ibid.
[19] Welty, "Old Mr. Marblehall," 114.

> wonder that filled my being, would arrest me and I could hardly bear my pleasure for my guilt.[20]

Communication technologies thus seem in Welty's work a largely passive medium not for communication so much as for the dissemination of one-way directives, logorrhea, orders, commands, information—a peremptory, intrusive one-way transfer of information, like Miss Duling's bell, and not amenable to the world or the human condition that Welty wanted to write about. All that would bring her world into the twentieth century, make it part of the larger world. *The Optimist's Daughter*, perhaps her most modern fiction, does indeed have jet planes in it, but it too is largely void of modern communications.

In Philip Roth's novel *Exit Ghost*, his doppelgänger, Nathan Zuckerman, returns to Manhattan after a ten-year idyll in the New England woods. His responses to the changes he discovers are much like Rip Van Winkle's upon awakening after twenty years of sleep. Of interest to us here are his observations about the current communicative phenomenon:

> What surprised me most my first few days walking around the city? The most obvious thing—the cell phones. We had no reception as yet up on my mountain, and down in Athena, where they do have it, I'd rarely see people striding the streets talking uninhibitedly into their phones. I remembered a New York when the only people walking up Broadway seemingly talking to themselves were crazy. What had happened in these ten years for there suddenly to be so much to say—so much pressing that it couldn't wait to be said? Everywhere I walked, somebody was approaching me talking on a phone and someone was behind me talking on a phone. Inside the cars, the drivers were on the phone. When I took a taxi, the cabbie was on the phone. For one who frequently went without talking to anybody for days at a time, I had to wonder what that had previously held them up had collapsed in people to make incessant talking into a telephone preferable to walking about under no one's surveillance, momentarily solitary, assimilating the streets through one's animal senses and thinking the myriad thoughts that the activities of a city inspire. For me it made the streets appear comic and the people ridiculous. And yet it seemed like a real tragedy, too. To eradicate the experience of separation must inevitably have a dramatic effect. What will the consequence be? You know you can reach the other person anytime, and if you can't, you get impatient—impatient and angry like a little stupid god. I understood that background silence had long been abolished from restaurants, elevators, and ballparks, but that the immense loneliness of human beings should produce this boundless longing to be heard, and the accompanying disregard for being

[20] Welty, *One Writer's Beginning*, 858–61.

> overheard—well, having lived largely in the era of the telephone booth, whose substantial folding doors could be tightly pulled shut, I was impressed by the conspicuousness of it all and found myself entertaining the idea for a story in which Manhattan has turned into a sinister collectivity where everyone is spying on everyone else, everyone being tracked by the person at the other end of his or her phone, even though, incessantly dialing one another from wherever they like in the great out of doors, the telephoners believe themselves to be experiencing the maximum freedom. I knew that merely by thinking up such a scenario I was at one with all the cranks who imagined, from the beginnings of industrialization, that the machine was the enemy of life. Still, I could not help it: I did not see how anyone could believe he was continuing to live a human existence by walking about talking into a phone for half his waking life. No, those gadgets did not promise to be a boon to promoting reflection among the general public.[21]

I suspect that Welty would have found this passage to her liking. I'm not sure it's fair to call her a Luddite, though I did ask her once in the early eighties whether she had yet bought herself a computer, and I remember the loathing and disdain—perhaps it was just despair—that passed over her face before she said, with her usual sweetness, "No, and I don't plan to." At any rate, her fictional attitude toward telephones seems to antedate Roth's animadversions against cellphones by several decades. More often than not, phones appear as tools of power, as when Miss Duling calls her former students to tell them how to run the city of Jackson; as instruments of illusion, as when the Weltys assume that her father saved her mother's life with champagne and a telephone call; and, like Victrolas, Dictaphones, and movies, as one-way communiqués, as with the caller who hardly lets her mother get in that edgewise word. In Welty, technology nearly always blocks communication rather than making it happen, although, to be sure, her human communicants don't always do much better face to face: so perhaps technology is primarily an extension of communication difficulties in general.

For Zuckerman, cellphones, those "gadgets[,] did not promise to be a boon to promoting reflection among the general public," and I do suspect that Welty's general sentiments about technology would be much closer to Roth's than to Bill Gates's; she would believe that they more often foster isolation than facilitate communication. The "growl" she invoked in contrasting her parents' love-duet to the sounds of the Victrola seems to stand for some suspicion of technology. That "growling" may be a little menacing, a threat latent with inarticulateness and misunderstanding. It's

[21] Roth, *Exit Ghost*, 64–65.

hard to forget that her father used the Victrola to supplant her mother in one of the most intimate of mother-child moments, lullabies at bedtime, apparently considering her lullaby voice less than perfect—and, as I suggested, we may well wonder whether her mother, who began their duet whistling, became self-conscious and switched to humming to keep from being criticized—or replaced—but in any case found herself, perhaps yet again, chafing under her husband's watchful and ever-so-nicely critical eye.

Victrolas distort the voices they reproduce, first in the growling as the turntable slowly turns to get started, then as it slows and has to be rewound, producing static and instability in the voices it reproduces: nobody who's heard one would confuse it with a real human voice. Welty's parents' duet distinctly favors her parents' joy, their love, which is different from the Victrola's "growling," which will replace her mother's voice. But in another oddly constructed sentence, she introduces that "*difference*" before telling us what her parents' duet is different *from*: the effect is more than a little disorienting, confusing. It's as if that growl has been lurking all along from the beginning of their marriage, as it is being wound up to begin, lurking there precisely to disrupt and confuse, to *clatter* into their pleasure in each other and, no doubt, in Eudora, who may personify that growl. The difference was the growl: that is, the Victrola growled "from the beginning," loudly and obviously. In her parents' duet, the growl is there, though not so obviously, hidden in the first flushes of love and family; but it is definitely *there,* lurking in the language that young Eudora hears and doesn't hear, in their complicated attempts at harmony.

That mechanical growl is thus a premonition that Welty imposes backward on what otherwise might be a completely happy memory. It's a premonition of her father's death, her mother's bitterness, her own career as a passionate observer of the ways human beings love and misunderstand each other, of all the clatter and growl even the happiest families make.

Works Cited

Marrs, Suzanne. *Eudora Welty: A Biography*. New York: Harcourt, 2005.

Prenshaw, Peggy Whitman. "The Construction of Confluence: The Female South and Eudora Welty's Art." In *The Late Novels of Eudora Welty*, edited by Jan Nordby Gretlund and Karl-Heinz Westarp, 176–94. Columbia: University of South Carolina Press, 1998.

Roth, Philip. *Exit Ghost*. Boston: Houghton Mifflin, 2007.

Waldron, Ann. *Eudora: A Writer's Life*. New York: Doubleday, 1998.

Welty, Eudora. "Magic." *Manuscript* 3 (September–October 1936): 3–7.

—. "Old Mr. Marblehall." In Welty, *Stories*, 111–19.

—. *One Writer's Beginnings*. In Welty, *Stories*, 831–948.

—. *The Ponder Heart*. In *Complete Novels*, 337–424. New York: Library of America, 1998.

—. *Stories, Essays, and Memoir*. New York: Library of America, 1998.

—. "Why I Live at the P.O." In Welty, *Stories*, 57–69.

CALLING THE ONE YOU LOVE: THE TELEPHONE IN MART CROWLEY'S *THE BOYS IN THE BAND* AND ROBERT CHESLEY'S *JERKER, OR THE HELPING HAND*

TOMASZ BASIUK

"You don't have to be gay to be a wanton."[1]

In a *trio* of essays on what he describes as borderline literary genres, Jan Potkański contends that the crossword, the personal ad, and the musical score are popular embodiments of the postmodernist preference for labyrinthine surfaces over the modernist fixation on the hidden and the profound.[2] The essay on personal ads, which I shall discuss, is specifically concerned with ads found in gay magazines that are often openly expressive of sexual desire. The argument about them is based on the Horatian *ut pictura poesis* principle, in that Potkański reads the inclusion of the subject's photograph in many ads as indicative of the genre's investment in the *literalness* of the communicative instance which it seeks to enact. Unlike the writing traditionally subsumed under the literary label, personal ads are intended to facilitate real-life contact between the author and the addressee. The inclusion of a photographic image invokes the ad placer's promised presence in a perhaps self-evident manner—barring such manipulative practices as publishing the photo of another man, or changing one's picture with technological tools—and thereby reinforces the prospect of actual sexual consummation should the author and reader come to such an understanding. That same picture, however—like, indeed, the meeting of bodies it anticipates—is also intelligible as text. The

[1] Spoken by Emory in Crowley, *Boys in the Band*, 111.
[2] Potkański, "Labirynt krzyżówki," 41–61; "Ciało w tekście," 153–92; and "Partytura jako rodzaj literacki," 337–72. The brief summary of an argument which follows reflects the content of the second of the three essays.

inclusion of the photograph in the ad makes it a part of the ad's text, and the sometimes visible manipulation of the image, such as a line drawn across the subject's eyes in an apparent, if not always successful, effort to disguise identity, serves no less to underscore the written quality of the image. The writing upon the body's photographic representation suggests that the body itself is to be regarded as a textual surface. Likewise, the sexual communication of bodies, promised in the ad, is less communion and more a combinatory practice of meeting and intersecting surfaces, presumably similar to the kind of encounter imagined with the term *intertextuality*.

There is, of course, a tension at work in these formulations, in that the argument is simultaneously for a reading that emphasizes the literal, i.e., actual *presence*, as the personal ad has first of all a utilitarian function in its promise to deliver as advertised, and for a reading emphasizing that the ad, along with the photo, is textual: an instance of writing with its attendant tropes. Potkański seems to gloss over this apparent contradiction because he is interested in unraveling another opposition: that between modernism and postmodernism, especially as it is construed in Polish studies, as well as in critiquing the modernist treatment of sexuality, especially in its non-normative variants, and specifically of male homosexuality. Simply put, he is attempting to topple the critical commonplace of connecting male homosexuality, construed as transgression, with figurations of death; a connection typically expressed as the Freudian duality of Eros/Thanatos. In so doing, he juxtaposes a range of modernist and contemporary writers (i.a. Iwaszkiewicz, Tkaczyszyn-Dycki, Musiał) whom he sees as followers of this modernist, and implicitly homophobic, conjunction with the sexually affirmative personal ad. One might parenthetically note that there appears to be a dearth of properly literary examples which might be employed for the purpose of such a juxtaposition, that being the presumed reason for turning to the personal ad. But then the ad is also heuristically useful because it illustrates so well the *real* possibility of sexual communication which a more narrowly literary example would merely invite the reader to imagine.[3] It is the direct appeal to sexual desire, in the absence of guilt and shame, that appeals about the ad, and that makes for such a nifty contradiction of the modernist sensibility. As Potkański points out, both defenders and detractors of sexually explicit modernist writing, whether by Joyce, Lawrence, or Zegadłowicz, oppose what they see as gratuitous

---

[3] I omit in this discussion Potkański's very interesting and highly amusing comparison of personal ads in gay magazines to the Hellenic erotic epigram, the priapea, and the Athenian Old Comedy; see Potkański, "Ciało w tekście," 177ff.

eroticism. However, while defenders view the erotic as an expression of something more elevated (for example, as evocative of social revolt) detractors' openly puritanical attitude leads them to recognize the erotic as the work's very point. Thus, the puritanically-minded censors treat the erotic more seriously, even if they "wish to ban that which postmodernism sets free."[4]

The celebration of the postmodern by Potkański has the ring of a manifesto belonging more properly to poetic rather than critical discourse. Perhaps the seeming absence of suitable literary examples accounts for the critic's assuming the heraldic role, as well as for the exquisitely ironic use of personal ads as the object of literary analysis. One can of course counter this celebratory tone with some well-rehearsed critiques of the postmodern: Slavoj Žižek's ventriloquized injunction to "enjoy your symptom!" as the capitalistic subject's ultimate guilt-trip comes to mind. Potkański, who discusses both Freud and Marx, is certainly aware of the paradox. In his essay on crosswords, he states that postmodern subjects feel comfortable in their labyrinthine prisons. While the essay on personal ads lacks a similar proviso to guard against the perhaps too easy investment in liberation through sex, the essays are cross-referenced and it is clear that they are informed by parallel insights.

The contradiction between the ad's literal meaning, dramatized by its utilitarian goal, and its status as *written* text, one which therefore relies on tropes (as evidenced by Potkański's comparison of the ads to classical poetic genres), is less explicitly theorized. While he freely admits that the ad is *written*, i.e. that it is a literary construction—an argument which he subsumes under the *ut pictura poesis* principle even while he sees the photo as also textual—Potkański insists that the personal ad is distinguished from most writing, especially that known as literary, by its expression of actual desire and the open invitation to undertake action

---

[4] "chcą zakazać tego, co postmodernizm wyzwala." Potkański, "Ciało w tekście," 192. Potkański's preference for the detractors' position is consistent with his valuation of literalness, evident also in the quoted phrase. However, I note in this context that his criticism of queer theory as aligned with forms of mourning that include the celebration of victim-status juxtaposes it with a progressivist model of modernization that is unproblematically affirmative of sexuality (see ibid., 187). Potkański's position is thus partly consistent with some criticisms of queer theory that arise from within it, notably Bersani's "Is the Rectum a Grave?" Haver's "Really Bad Infinities," and Ricco's *Logic of the Lure*, but Potkański is considerably more optimistic about the liberating potential of sex than these other writers. For a defense of mourning as crucial to political activism, see Crimp's *Melancholia and Moralism.*

other than the act of reading. This argument about desire is open to the same criticism as the celebration of the postmodernist "setting free" of enjoyment: the Freudian and the Marxian approaches, invoked in his essays, have of course been used to show that desire itself can be an effect, as well as a means, of socialization, rather than reflect the subject's autonomous will, or his or her nature. Therefore, I contend that what distinguishes the ad in Potkański's argument is its *literalness* rather than the presence of desire (including the literal way in which desire is understood to be present in the personal ad). It is the literalness of desire that is supposed to be set free, rather than desire as such.

Lee Edelman has offered an especially compelling critique of the discourse of literalness in connection to homosexuality. His argument includes a reading of Derrida's "Plato's Pharmacy," a seminal critique of the notion that speech, unlike writing, is a bearer of directly available meaning. This belief about speech is seen as participating in a metaphysics of presence, itself a mythicizing view of language. Following Derrida's point, Edelman argues that the supervaluation of literal meaning to the detriment of tropes and textuality indirectly expresses a preference for a reproductive sexuality that results in childbearing over non-reproductive sexualities coded as barren.[5] In other words, an investment in the literal necessarily involves a return of the repressed modernist coupling of Eros and Thanatos whenever non-reproductive sex is concerned.

With Edelman's point in mind, my own reading of two gay plays will forego the premise of *ut pictura poesis* deployed by Potkański in order to focus on the opposition between writing and speech, also as elaborated in the above-cited argument. Mart Crowley's *The Boys in the Band* (1968) and Robert Chesley's *Jerker, or The Helping Hand* (1985) both deflect speech exchanged between people involved in amorous encounters—including speech literally exchanged between characters on-stage—with the use of the telephone. Speech, aka *phone* or *logos*, is made distant to itself when it is sent a distance with the use of a technical device. This self-distancing makes speech similar to writing in the Derridian sense. When *phone* becomes *tele-phone*, it undergoes a deconstructive procedure in which its seemingly unequivocal and immediately available meaning turns out to be dependent on various circumstantial determinants. A confession of love on the telephone is not quite identical to a confession of love executed in the direct presence of the lover, nor is it heard the same

[5] Edelman makes this argument about the ubiquitous "Silence=Death" slogan of AIDS activism, devised by ACT-UP. Edelman, "The Plague of Discourse: Politics, Literary Theory and 'AIDS,'" in *Homographesis: Essays in Gay Literary and Cultural Theory*, 79–92.

way as when the lover uttering the amorous statement is actually there for us. Distance between lovers qualifies such expression, and possibly undermines it. The distance of space is somehow also emotional distance, a gap created by external circumstances that nevertheless impacts the lovers' relationship, making it less something that is simply "given" and more a thing that is posited, perhaps as challenge, and, by the same token, a thing that can go wrong, fail altogether, or turn out differently from what one supposed.

The Boys in the Band, which opened in New York City, where it is set, a year before the Stonewall Riots of 1969, seems to announce the imminent breaking down of the closet doors even as it investigates the internalized homophobia of the play's gay male characters, caught in their versions of the Eros=Thanatos precept. The form which this equation takes is informed by the received psychoanalytical views of the era, so that characters exemplify instances of arrested development, passive-aggressive behavior, chronic depression, etc., and occasionally indicate that these emotional difficulties are related to their homosexuality. Thankfully, there is also a more liberationist strain in the play, especially when the characters act out some of their desires—particularly through camping it up—without heeding the social restrictions on homosexual expression. Still, this inevitably causes tension, and the way that the play pre-figures the queer sexual rebellion and, indeed, the Stonewall riots, is as a social gathering that erupts in an emotionally violent clash. An invocation of the Stonewall riots is perhaps intended in the 1970 film version, which portrays the usual chaos of dinnerware and party things left behind by the now-departed guests as nearly overwhelming wreckage: "Do you think there's any possibility of just burning this room?"—Michael asks.[6] In the course of the evening, Michael, the host of a party celebrating his friend Harold's birthday, to which only a circle of gay male friends have been invited (one of whom brings a hot male trick as a birthday present)—and which rapidly deteriorates into a nightmarish free-for-all somewhat similar to Edward Albee's *Who's Afraid of Virginia Woolf?*—challenges his guests to call the person whom they have always loved in order to confess their feelings for the first time. This predictably turns out to be an embarrassing experience for those who venture to do so, especially if they take the host's challenge literally, like the Afro-American character Bernard, who attempts to call the son of his mother's former employer, the scion of a southern family with whom he had shared a drunken homosexual experience some years ago. The phone is answered

---

[6] Crowley, *Boys in the Band*, 130.

by the white man's mother and the possibility of a direct confession is prevented, as Bernard is overcome with embarrassment and with the wine he has been imbibing to numb his sense of shame. Only Alan, Michael's drunk and violent former college classmate, ostensibly straight, who has walked in uninvited, uses the occasion of the game to make up with his estranged wife, frustrating attempts by others to uncover his alleged homosexual yearnings. Everyone else seems to find the ordeal deeply shaming and usually unrewarding.

In one instance, two characters call each other. Hank and Larry have been lovers and have lived together for some time though Hank is married (and is in the process of divorcing his wife), and though Larry insists on his right to be promiscuous while in a relationship with Hank. When Hank's turn to place a call comes, he dials their answering service and leaves a message of love for Larry. He is in fact speaking to an operator who knows them both (or at least knows their voices and identities), so that Hank's confession of love is also a coming-out. Larry reciprocates by calling Hank on an extension in Michael's apartment and, speaking to his lover, who is in another room, confesses his love—". . . For what it's worth."[7] Larry's statement is qualified further by the revelation that he and Donald, one of the party guests, have had anonymous sex in the past, and by an earlier exchange between Larry and Hank, in which Larry refuses to abandon his promiscuous habits and accuses Hank of fabricating a pact which allegedly allows a *menage* but disallows sexual partners outside the relationship. As these exchanges between Hank and Larry illustrate, the meaning of the words "I love you," while perhaps not vacuous, is uneasily pinned down, and irreducible to just one sense. The technical intervention of the telephone used as dramatic device serves to displace the presumptive directness of speech. It casts the answering service operator as a witness to, and in fact, the literal recipient of, Hank's confession of love, deflecting this intimate confession into public space. Moreover, Hank and Larry's communication via the telephone seems to open up the space for Larry's numerous provisos to his confession of love. Thus, the presumed directness with which speech proceeds from sender to receiver and the presumed clarity of the speaker's intention are easily falsified. Instead, multilayered meanings come across, endowing speech with the quality of writing.

The one instance of seemingly flawless communication is the conversation in which Alan makes up with his wife. This confirms Edelman's point that *literal* language, i.e., language characterized by the

---

[7] Ibid., 117.

directness of speech, and seemingly liberated from the vagaries of writing, is the stereotypical prerogative of reproductive heterosexuality. Of course, the play in fact contextualizes this particular conversation by putting in doubt the veracity of Alan's heterosexual commitment ("Why do you think he stayed"—Donald asks Michael as the evening comes to an end).[8] More crucially, it dramatizes the comparative difficulty of homosexual affective expression. It is to Crowley's credit that he was able to trace, in the exchanges between his characters, the complications of what it might mean to be gay in a manner that has on-going significance, including interstitial meanings created by the meeting of sexuality and race, sexuality and class, etc. His characters' confessions go further than affirming one's right to have erotic feelings for members of one's sex.

This elaboration is accomplished with the participation of witnesses, as none of the conversations is private: it is, after all, a party game that everyone is playing, and all statements are made, in a manner of speaking, on a party line. Except for the presumed heterosexual Alan, everyone is breaking some taboo about sexuality, e.g. by admitting his preference for a same-sex partner to a third party, by confessing to a cathexis that breeds contempt, or by insisting on his right to be promiscuous while in a relationship. Moreover, these transgressive confessions are publicly performed, so that the sense of shame they bring on forms a volatile bond among members of the clique. Alan, meanwhile, admits to nothing. His call to his wife is an exemplary deployment of the pharisean injunction to see no evil, hear no evil, speak no evil. His gay best friend never learns the reason for the marital crisis, or the exact nature of the resolution that Alan seems to have made in the course of the evening. In other words, Alan's statement to his wife (and to the men who are listening to his words) is necessarily a palimpsest of meanings which, however, he manages to keep hidden. His form of addressing his wife and, simultaneously, the gay men at the party, is a masquerade in which he passes for whatever constitutes normal. Curiously, this has the effect of putting the equation of Eros and Thanatos on its head, as it is the straight male's speech whose actual meaning disappears down a long labyrinth of pretence while the gay characters have dared to make their own dissembling strategies momentarily visible. The play's mode of presenting homosexual characters onstage is simultaneously a mode of their dis-appearing or, to use José Munoz's term, dis-identification: their moment of exposure renders visible the process of constructing an identity rather than puts forth a snapshot suggestive of a ready-made product.

---

[8] Ibid., 130.

Jerker, or The Helping Hand is set in San Francisco a few years after the onset of AIDS. While *The Boys* pre-figured the post-Stonewall era, defined by the identity politics sometimes dubbed "gay pride," *Jerker* documents a period in which the visibility of gay males (among other so-called "risk groups") was radically enhanced and at the same time reconfigured through the *Totentanz* of the epidemic. Bert, a gay man, gets a call from J.R., a trick about whom he had forgotten. The call is sexual, and they develop a quasi-anonymous sexual relationship which the play frames with an anecdote about two men who have carried on a sexual relationship for fifteen years without exchanging a word except to make the most rudimentary arrangements for their meetings. The anecdote is reminiscent of the 2001 film *Intimacy*, which, though concerned with a long-lasting anonymous sexual relationship between a woman and a man, was co-written and directed by the gay director Patrice Chéreau. In *Jerker*, the relationship is carried out entirely with words spoken on the telephone, a circumstance which dramatizes the eighties' preoccupation with the exchange of body fluids and the exigencies of safer sex. There are several occasions in the play for characters to reminisce about the times before the epidemic. The play ends when one of the men can no longer be reached, his phone having been disconnected. His friend and the audience assume he has died, possibly of AIDS.

While the last message which J.R. leaves on Bert's phone before it is disconnected includes the words "I love you," their conversations have already made it clear that the two men share a connection that they feel comfortable describing as affectionate and intimate. These words, like indeed the word "love," are not used in a manner that would render them distinct from talk about sex. On the contrary, Bert describes an anonymous sexual experience as love.[9] Bert and J.R.'s love is not modeled on the heterosexual couple, and neither is their sex. The men's conversations are forms of writing, and involve leaving recorded messages for one another. Each man is also imagining the other's body and inventing fantasies that they can share. Their intimate exchanges are an apt illustration of the investment which I understand Potkański to be making in the desirability of a direct presence of desire that postmodernism is supposed to have enabled. However, it is clear that these intimate exchanges are a form of writing. The two men are not *literally* having sex, if having sex literally involves bodily contact: their entire relationship is, in an important way, invented and imagined, as well as mediated by the telephone, though not

[9] He remembers it as sex with the first man he had ever seen wearing a walkman (in the 1970s). Chesley, "Jerker, or The Helping Hand," 481–3.

any less real for all that. This is perhaps not surprising. Potkański cites Thomas Nagel to the effect that sexual excitement is incited by the reactions one observes in the other to the reactions that the other is causing in oneself. This interactive model of sexual desire requires a sort of quantum leap to be initiated. It is not a model of simple reciprocity but of reciprocal inventiveness, a model which involves writing the other's body. This imaginative writing is enacted on the stage in the imaginary scene in which J.R. crosses to Bert's bedside, while Bert continues to speak into the phone.

Another way to put this is that Bert and J.R.s' relationship is mediated by the communication technology which they are using as a sexual toy that comes in handy because of the health risk involved in bodily contact, and apparently because one of the men is a handicapped Vietnam war veteran, hesitant to disclose this circumstance. The use of the toy in no way diminishes the sex, nor takes away from its authenticity, unless one adopts the restrictive view of sex in which the reproductively oriented heterosexual variant is the fixed norm. While the phone has the effect of rendering one man invisible to the other (J.R. remembers what Bert looks like), and of putting their sex out of sight, it is no less explicit for that. Indeed, the play seems to toy with the concepts of "obscenity" and "putting-on-stage" through its deployment of sexually charged speech coupled with a refusal to have characters enact the sexual acts they describe. As in *The Boys*' appearing through a disappearing act, there is in *Jerker* the work of making visible precisely through a configuration of invisibility. As with the previously discussed play, I read this seeming paradox as pointing to the vagaries of writing, in contradistinction to the presumed directness of speech. It is the process of making happen that is brought to the fore, rather than the consummation to which it leads.

In an essay titled "Homographesis," Edelman speculates about the existence of the gay body. His argument centers on the relationship between metaphor and metonymy: recognizing a gay body requires a metaphoric leap the taking of which enables one to read the signs of gayness: metonymies connect into larger wholes governed by the metaphor. The ontological question this raises is whether metaphor determines the metonymic chains of evidence, or the other way round.[10] The figure of synecdoche, and the pictorial and narrative convention of *mise-en-abyme*, dramatize the dual movement between metaphor and metonymy. The part which can stand for the whole functions as both

[10] Edelman, "Homographesis," in *Homographesis: Essays in Gay Literary and Cultural Theory*, 3–23.

metonymy and metaphor, and therefore embodies the duality. In both plays, talking on the telephone functions as a dramatic device comparable to the *mise-en-abyme* because the telephone simultaneously promises to deliver the interlocutor's voice as speech pure and simple, and turns that voice into a form of writing by introducing a gap between the voice and the speaker. This, of course, is a self-representation of the theater, which deploys writing through speech, and presents speech as though it were writing.

What is at issue in both plays are the forms that male same-sex relations can take given their socially marked difference from the reproductive/heterosexual norm, and under conditions that range from (internalized) homophobia to the AIDS epidemic. Also at stake are the modes of self-expression, or perhaps self-exposure, that these men are employing vis-à-vis one another and in the face of on-lookers: specifically in the *The Boys*, but also in *Jerker* through its strategy of teasing the audience with lurid speech acts and simultaneously withholding the satisfaction of corresponding visual representation. What is also at issue are the forms of relationality which these plays are putting to the audience: a proto-liberationist stance projected (successfully, as the Stonewall riots and subsequent events confirm) from the sixties' sexual revolution, and an affirmation of anonymous sex and promiscuity in the face of the AIDS epidemic. Unlike the personal ads, these plays do not offer the prospect of sexual communication with their authors. (Though one may not, of course, preclude this outcome. One may, moreover, be reasonably certain that both plays provoked sexual encounters between members of the audience, the cast, etc.) However, it would be misleading to say that representations of the erotic in these plays are subservient to some higher goal. *Jerker* is subtitled "A Pornographic Elegy with Redeeming Social Value and a Hymn to Queer Men of San Francisco in Twenty Telephone Calls, Many of Them Dirty," but the irony in the subtitle is as clear as in the epigraph from Walter de la Mare which ends with "Not two of us, but three." Similarly, the line from *The Boys* I picked for my epigraph—"You don't have to be gay to be a wanton,"—hardly implies any criticism of sexual exuberance, whether gay or straight.

While it may be argued, I believe correctly, that these plays are invested in the political project of gay community building, it is just as accurate to say that this project's goal is to enable sexual encounters, including, specifically, those that are non-reproductive, non-heterosexual, and anonymous, as is evidenced in some of Larry's statements in *The Boys*. If the very meaning of "community" is challenged by this formulation, and by the project which the term is brought to describe, so be it. The same project is, of course, found in innumerable personal ads.

Works Cited

Bersani, Leo. "Is the Rectum a Grave?" In *AIDS: Cultural Analysis/Cultural Activism*, an October Book, edited by Douglas Crimp. Cambridge, MA: MIT Press, 1988.

Chéreau, Patrice. *Intimacy*. DVD. France, UK, Germany, Spain: 2001.

Chesley, Robert. "Jerker, or The Helping Hand." In *Out Front: Contemporary Gay and Lesbian Plays*, edited by Don Shewey. New York: Grove Press, 1988.

Crimp, Douglas. *Melancholia and Moralism: Essays on AIDS and Queer Politics.* Cambridge, MA: MIT Press, 2002.

Crowley, Mart. *The Band Plays: "The Boys in the Band" and "The Men from the Boys."* With an introduction by Donald Spoto. Los Angeles: Alyson Books, 2003.

Edelman, Lee. *Homographesis: Essays in Gay Literary and Cultural Theory*. New York: Routledge, 1994.

Haver, William. "Really Bad Infinitives: Queer's Honor and the Pornographic Life." *Parallax* 5, no. 4 (October 1999): 9–21.

Munoz, José Esteban. *Identifications: Queers of Color and the Performance of Politics*. Minneapolis: University of Minnesota Press, 1999.

Potkański, Jan. "Ciało w tekście." In Potkański and Mrugalski, *Literackie obrazy nieobecności*, 153–92.

—. "Labirynt krzyżówki." In Potkański and Mrugalski, *Literackie obrazy nieobecności*, 41–61.

—. "Partytura jako rodzaj literacki." In Potkański and Mrugalski, *Literackie obrazy nieobecności*, 337–72.

Potkański, Jan, and Michał Mrugalski, eds. *Literackie obrazy nieobecności*. Warszawa: Wydział Polonistyki Uniwersytetu Warszawskiego, 2005.

Ricco, John-Paul. *The Logic of the Lure.* Chicago: The University of Chicago Press, 2002.

Žižek, Slavoj. *Enjoy Your Symptom! Jacques Lacan in Hollywood and Out*. New York: Routledge, 1992.

Home Calling / Calling Home: Women and Telephony in the Works of Chitra Banerjee Divakaruni and Amulya Malladi

Izabella Kimak

Women making phone calls to and receiving phone calls from India figure prominently in the literary output of the South Asian American diaspora. In the present paper I intend to analyze selected novels and short stories by Chitra Banerjee Divakaruni and Amulya Malladi to map the roles telephony plays for those female characters who are first-generation immigrants from India. First, I plan to discuss the practice of telephoning as an attempt at exerting control upon women's lives. Frequently, India-to-America phone calls are meant to ensure that the connection between the homeland and the new land is maintained and that women far from home still adhere to the rules of their native culture. On the other hand, making phone calls is also a process through which alliances are forged between women who reach out to one another across the Pacific to help in times of need. Those two patterns, present in Divakaruni's texts, are modified to some extent in Malladi's *The Mango Season*, as we shall see later. This paper also aims to explore the intersection between women's call-making and the notion of home.

In order to explore the first aspect of telephoning mentioned above, that is controlling through phone calls, I would like to focus on Divakaruni's short story "The Word Love." The story, which comes from the writer's first collection titled *Arranged Marriage*, published in 1995, is written in the second person singular. The unnamed protagonist is a young Indian woman who has come to the United States to study for her PhD degree in literature. For her, telephone calls are simply a means of keeping in touch with her mother back in Calcutta, and in this way of remembering her roots. The relationship between the two women is by all means a difficult one. What the heroine feels for her mother is a mixture of love and fear. She acknowledges the similarities between herself and her mother as well as her mother's dedication "to the business of bringing [her] up" on her own.[1] However, for the main character, the mother's primary attributes are her sternness and strictness. In her eyes, her mother is a guardian of tradition and a paragon of dutifulness. Thus, even as an adult living thousands of miles away from her hometown, the protagonist cannot help feeling guilty when she goes against any of the values inculcated in her by her mother. The experience of studying abroad has naturally influenced her in many respects and has forced her to rethink her preconceived attitudes, yet she strives to keep those changes a secret so as not to upset her conservative mother. She is also aware of the dire consequences of overstepping the boundaries, as she remembers what her mother used to say: "Better no daughter than a disobedient one, a shame to the family."[2]

As the heroine and narrator of Malladi's novel *The Mango Season* rather humorously puts it, there are certain types of behavior that the parents of children headed for North America admonish them to steer clear of. The orders that the parents give are as follows:

> Do not eat beef. (The sacred cow is your mother!)
> Do not get too friendly with foreign people; you cannot trust them. Remember what the English did to us.
> Cook at home; there is no reason to eat out and waste money.
> Save money.
> Save money.
> Save money.
> DO NOT FIND YOURSELF SOME FOREIGN MAN/WOMAN TO MARRY.[3]

The last item on the list is by far the most crucial, which fact is emphasized visually in the book by the use of capital letters, and it is precisely the one that Divakaruni's heroine has broken. The protagonist knows that her behavior stands in marked contrast to what is expected of a young woman of her class/caste in India: to guard her virginity until a suitable groom is found and her marriage is arranged by her family. According to the sociologist Jyoti Puri, who studies female sexuality in its contemporary middle-class Indian context, "marriage and motherhood are considered primary gender roles for women across social classes."[4] As many as 95 percent of marriages in India are arranged and "a predominant

---

[1] Divakaruni, "Word Love," 58.
[2] Ibid., 62.
[3] Malladi, *Mango Season*, 3.
[4] Puri, *Woman, Body, Desire*, 136.

expectation pertaining to middle-class women's sexualities is pre-marital chastity."[5] Thus, even as love propels Divakaruni's heroine to inform her mother that she is living with an American boyfriend, the fear of the older woman's reaction prevents her from doing so. When she moves in with her boyfriend, she makes sure that she is the one to pick up the phone, not only on Saturday mornings, when it is her mother's habit to call her, but any time the phone rings. At the same time, she plans to disclose everything and she even practices the words which she is going to use. However, when the time comes, "[her] heart flings itself around in [her] chest like a netted bird, and the words that [she] practiced so long are gone."[6] This image of a captured bird eloquently expresses the power relation in the Indian mother-daughter interactions.

When the mother eventually discovers that her daughter is in a relationship with an American, her reaction is exactly as the heroine has predicted; according to the mother, rejection and disowning are a suitable punishment for her daughter's disobedience. Hence, the mother-daughter relationship is severed despite the heroine's frantic attempts to seek forgiveness. When the daughter recovers from the trauma, she begins to understand how stifling her relationships both with her mother and her boyfriend are. On the verge of committing suicide, the protagonist finally decides to shed the burden of trying to please others that she has so long borne and to start living only for herself.

Telephone conversations in "The Word Love" are thus a means through which the mother tries to retain control of her daughter's life, especially of her sexual behavior. The daughter, however, subverts these attempts by always answering the phone herself and keeping it secret that she "is living in sin."[7] The telephone line is like the umbilical cord that connects the mother to the daughter. Ironically, the umbilical cord is cut by the mother, not by the protagonist as her boyfriend has repeatedly advised her to do, when the older woman rejects her daughter altogether and her phone goes dead. As has already been pointed out, this experience turns out to be a blessing in disguise as it helps the protagonist to mature.

In Divakaruni's novel *Sister of My Heart*, published in 1999, telephone conversations are used by female characters to strengthen the connection between women on both sides of the Pacific Ocean and to offer assistance to one another. The book portrays a close relationship between two young women, Anju and Sudha, brought up in the same household by their widowed mothers. Their bond is weakened when they are married off;

---

[5] Ibid., 109, 138.
[6] Divakaruni, "Word Love," 60.
[7] Ibid., 59.

Anju accompanies her husband to America, whereas Sudha stays in India and lives in a joint household with her in-laws. The two childhood friends find themselves in totally different circumstances. Anju goes to college in America to study literature while Sudha fulfills all the traditional duties of a young wife in India: she takes care of her husband and his family and has to obey her strong-willed mother-in-law.

The different ends to which Anju and Sudha use their telephone conversations together with the influence of the geographical setting upon their lives are brought to the fore a few years later when the two women become pregnant. They both undergo amniocentesis to make sure the babies do not have a genetic defect. When it turns out that Sudha is expecting a baby girl, her mother-in-law pressures her to have an abortion as she cannot accept a girl as the first grandchild in the family. Her own pregnancy makes Anju identify with Sudha and, aware of the fact that her cousin cannot depend on any aid from her spineless husband or her own mother, for whom notions of propriety and respectability demand that Sudha stays married, she takes it upon herself to help her cousin out. Hence, Anju deploys the telephone line in her crusade to encourage Sudha to fight against the practice of female fetus abortion, which has been relatively common amongst upper-caste families in north India, for whom having a male heir is still a must.[8]

Sudha reciprocates Anju's involvement when it is the latter who needs emotional support after a miscarriage. To help her cousin deal with the trauma Sudha tells her a story about the queen who has to flee her kingdom when she is discovered to be carrying a baby girl. When Sudha stumbles, not knowing how to finish her story, Anju takes over, saying:

> She [the queen] saw a rainbow that extended all the way from the other side of the earth to her. . . . [T]he queen had forgotten that she had a twin sister who lived in the land across the ocean. The sister was sending her all her love in the form of this rainbow. . . . And her sister pulled her across the ocean, over the gaping jaws of sea monsters, to safety.[9]

The story, which takes its shape in the process of being told, emphasizes the strength which a relationship with a female friend can offer, although in the novel it is actually the "queen's sister," that is Anju, who is "saved" through the healing power of the story. In fact, the telephone connection is like the rainbow in the story via which Sudha helps Anju come to terms with her loss. The two women's use of telephone conversations is an

---

[8] Cf. Mandelbaum, *Women's Seclusion*, 65.
[9] Divakaruni, *Sister of My Heart*, 287–88.

example of how female characters resort to technology to create networks of female friendship and support when direct, face-to-face contact is not an option.

As Andre Aciman has once put it, "[a]n exile is not just someone who has lost his home; it is someone who can't find another; who can't think of another."[10] The elusiveness of the concept of home is stressed in all the three texts discussed in this paper, but perhaps the most so in Amulya Malladi's *The Mango Season*. The protagonist and narrator of the novel, Priya, after spending seven years studying and working in the United States, makes a visit to her hometown Hyderabad in South India to notify her family that she is engaged to be married to an American. The journey home is also important for another reason: Priya goes back to her country only to find out that it is totally unlike the India she remembers. She is astounded to realize that she no longer feels at home in her home town. She voices her feelings towards India thus: "I was seeing this world, my ex-world from my Americanized vision. This ex-world of mine was different to me now from what it had been before. I saw some things better, while other things had blurred beyond recognition."[11] Priya consistently uses the word "home" to refer to America, for example when she says:

> This [India] was not home anymore. Home was in San Francisco with Nick. Home was Whole Foods grocery store and fast food at KFC. Home was Pier 1 and Wal-Mart. Home was 7-Eleven and Starbucks. Home was familiar, Hyderabad was a stranger; India was as alien, exasperating, and sometimes exotic to me as it would be to a foreigner.[12]

On a conscious level, Priya seems positive that India is no longer her home, that the experience of emigration has transformed her to such an extent that she no longer fits there. However, at times the word "home" slips out almost accidentally when Priya talks about India, which perhaps proves her confusion about where she belongs. This may also bear out Aciman's observation noted above that an exile, or an emigrant for that matter, is never capable of finding another home having lost her original one. Aciman argues that exiles see and feel double, so presumably the word "home" simultaneously evokes the images of both the original home and the acquired one.

---

[10] Aciman, "Shadow Cities," 21.
[11] Malladi, *Mango Season*, 96.
[12] Ibid., 134.

As for telephone conversations, two types are present in *The Mango Season*. First, after her arrival in the States, Priya calls her family in India every weekend; then, when she is already in India, she phones her boyfriend in America instead. The first category of phone calls is akin to the one presented in Divakaruni's "The Word Love" in that Priya's kin make use of the opportunity to warn her against Westernization (an example of which is her having her hair cut) or against disobeying her family (for instance by choosing a love marriage, as one of her cousins in India did). In this way the patriarchal family wants to make sure that its members, especially the women, will stick to the family mores regardless of the place where they live or their own attitudes towards the mores in question. Priya, in turn, deploys the telephone connection together with email to achieve support from her partner when she finds out that her family have arranged a bride-viewing ceremony and are planning to marry her off. When she is unable to get through to Nick she panics, thinking that he is disgruntled with the fact that she has still not informed her family about their matrimonial plans. It seems, therefore, that for Priya the telephone line is a link to her American home, and when this connection fails, in Priya's eyes her whole life in America seems jeopardized as well.

In his 1993 essay "Immigrants and Family Values" Francis Fukuyama, reflecting on the perishing of conservative family structures among white Americans, asserts that immigrants from the so-called Third World tend to "have stronger family values than white, middle-class, suburban Americans."[13] Over a decade later, the same idea is reiterated in another essay of his, titled "Identity and Migration," in which he argues that "first-generation immigrants have usually not made a psychological break with the culture of their land of birth and carry traditional practices with them to their new homes," including rigid family structures. Why then does the psychological condition of the first-generation immigrant delineated by Fukuyama not seem to apply to Divakaruni's and Malladi's fictional characters? Why do those female protagonists, recent arrivals to the US, not think with nostalgia of their lost homelands and—instead of planting Indian-ness on the American soil, as Fukuyama would have it—are more likely to plunge headlong into the American lifestyle? Perhaps the answer is that being a woman puts one in a far more disadvantaged position in India than in America. Without idealizing America as the Promised Land, those fictional female characters, and their authors for that matter, realize that the US can offer them an escape from the stringent patriarchal world of their culture. It is not insignificant that both Divakaruni and Malladi

---

[13] Fukuyama, "Immigrants and Family Values," 157.

share an upper-class background, which for Indian women has traditionally been tantamount to confinement to the domestic sphere. If Fukuyama's observations hold water, it is only in reference to Indian men, who, as privileged members of traditional families, have reason to maintain the status quo after emigration. Personally, I would rather agree with Sucheta Mazumdar, who, in the introduction to *Making Waves: An Anthology of Writings By and About Asian American Women,* argues that "[f]or immigrant women arrival in America can be liberating" as "[s]ocietal norms of the majority community frequently provide greater personal freedom than permitted in Asian societies."[14] Divakaruni's and Malladi's protagonists exist outside the realm of the male-dominated extended family system, as two of them are on their own in America, whilst Anju lives only with her America-educated husband. Thus, without rejecting their cultural roots, they can make use of the opportunities available to them in America, opportunities of which their Indian society would rather deprive them. Even if, as Aciman puts it, they will never feel truly at home in America, it is certain that they do not mourn the loss of their Indian homes. And when they pick up the phone to talk to a friend or a relative, it is to offer or seek support, or to subvert the endeavors aimed at limiting their independence.

Works Cited

Aciman, Andre, ed. "Shadow Cities." In *Letters of Transit: Reflections on Exile, Identity, Language and Loss*. New York: New Press, 1999.

Divakaruni, Chitra Banerjee. "The Word Love." In *Arranged Marriage*. New York: Anchor Books, 1999.

—. *Sister of My Heart*. New York: Anchor Books, 1999.

Fukuyama, Francis. "Identity and Migration." *Prospect* 131 (February 2007). http://www.prospect-magazine.co.uk/article_details.php?id=8239 (accessed March 11, 2009).

—. "Immigrants and Family Values." In *Arguing Immigration: The Debate over the Changing Face of America*, edited by Nicolaus Mills. New York: Touchstone, 1994.

Malladi, Amulya. *The Mango Season*. New York: Ballantine Books, 2003.

Mandelbaum, David G. *Women's Seclusion and Men's Honor: Sex Roles in North India, Bangladesh, and Pakistan.* Tucson: University of Arizona Press, 1993.

[14] Mazumdar, "General Introduction," 15.

Mazumdar, Sucheta. "General Introduction: A Woman-Centered Perspective on Asian American History." In *Making Waves: An Anthology of Writings by and about Asian American Women*, edited by Asian Women United of California. Boston: Beacon Press, 1989.

Puri, Jyoti. *Woman, Body, Desire in Post-colonial India: Narratives of Gender and Sexuality.* New York: Routledge, 1999.

THE GRAPHIC REVOLUTION AND REPRESENTATIONS OF AFRICAN AMERICANS IN AMERICAN MAGAZINES, 1880–1900[1]

MATTHEW WILSON

In this essay, I am going to take a commonplace about the history of magazine technology at the end of the nineteenth century and use that commonplace to read a fragment of the history of illustrations of African Americans in elite magazines such as *Harper's* and *Scribner's*. The commonplace in the history of illustrations is this: the end of the nineteenth century was "a period of accelerated visual transformation in American culture. . . . Phrases such as 'graphic revolution,' 'iconographical revolution,' and 'magazine revolution' describe the changing role of visual images in print during the late 1800s and early 1900s."[2] As Luther Mott has pointed out in his major survey, *A History of American Magazines*, there was an explosion in the numbers of magazines toward the turn of the century. In the years 1885 to 1890, the total number of magazines increased by over a thousand to 4,400, and by the year 1900, there was a total of over 5,500 magazines.[3] Obviously, there was, as Mott observes, "a vast new audience for the American general magazine" in the 1890s.[4]

Several technological developments enabled this "graphic revolution." The first was the development of "high speed rotary presses"[5] and a drop in the price of paper, which allowed magazines to be sold in numbers never seen before. (For instance, the *Ladies Home Journal* was selling seven hundred thousand copies an issue in 1893.) Thus, magazines were

---

[1] I would like to thank my colleague at Penn State Harrisburg, Robin Veder, for her helpful feedback on this essay.

[2] Brown, "Popular Art."

[3] Mott, *History of American Magazines*, 11.

[4] Ibid., 8.

[5] *Society of Illustrators* Web site, "The History of Illustration: 1880–1890," http://www.societyillustrators.org/index.cms (accessed March 24, 2009).

ubiquitous at the end of the century, but the major, parallel development that was responsible for this "graphic revolution" was the invention of the "new, cheap technique of photoengraving known as the halftone."[6] "This revolutionary process" translated "tonal pictures into a pattern of tiny, varied dots . . . The new engravings could be made more quickly and were far less costly. Publishers were therefore able to increase the number and quality of pictures."[7] Previously, for instance, the *Century* magazine had "paid up to three hundred dollars for a page-size woodcut," whereas a halftone could be bought "for less than twenty dollars."[8]

Critics have also argued, however, that this change was more than simply a technological one; it changed how viewers perceived images. For instance, Estelle Jussim has argued that

> the heart of the iconographical shift was the disappearance of the "sign function" of the reproduction; it became "an optical sign with surrogate power." Previously . . . readers encountering illustrations were made aware of the subjective, contrived character of the pictures; these were obviously the product of an artist's mind and tools. And they still were, when a halftone reproduced a painting or a print. But with the halftone photograph, a photomechanical image, the illusion of seeing an actual scene, or receiving an objective record of such a scene, was immeasurably enhanced.[9]

I would like to argue explicitly with Jussim in this instance: I agree with her argument that before the advent of the halftone readers were aware of the "constructedness" of the image, but I want to entertain the idea that when illustrations began to be reproduced in halftones, the images were seen as less "constructed" and more as reproductions of "an objective record," a development that may have helped to increase the objectification of African Americans in the period and that then became one important element in the constellation of racist discourse in the United States that resulted in an unprecedented upsurge of racism in the United States in these years.

To begin with typical illustrations before the advent of the halftone, I want to look at several examples from a humorous sketch published in 1880 by *Harper's*, one of the preeminent literary journals of its day. It is by Sherwood Bonner entitled "Hieronymus Pop and the Baby," and illustrated by Arthur B. Frost; the story was reprinted with Frost's illustrations in Bonner's 1883 volume, *Dialect Tales*, published by *Harper's*

---

[6] Mott, *History of American Magazines*, 5.
[7] *Society of Illustrators*, "History of Illustration" (see note 5).
[8] Mott, *History of American Magazines*, 5.
[9] Jussim, *Visual Communication*, 206–7.

Fig. 1. Illustration by A. B. Frost, "Hieronymus sings a soothing ditty," *Harper's New Monthly Magazine* 61, no. 361 (June 1880), p. 21

the most prestigious publisher at the time. Luther Mott has said that the illustrator, Arthur B. Frost, "amused a whole generation of readers by his humorous delineations of rural characters" and achieved the "highest rank in illustrations of Negro life."[10] The tone of the story is set by an African American family that goes off, for entertainment, to see a hanging. The son, Hieronymus, has to stay home and take care of the baby "while his mother and the rest of them trotted off, gay as larks, to see a man hanged" (fig. 1).[11]

The boy's name alone alerts the reader that she is in typical generic territory: the ludicrous and inappropriate name, "Hieronymus," signals that nothing in the story is to be taken seriously because it belongs to the long American tradition of seeing African Americans as simple and humorous. In terms of Jussim's argument, though, Frost's woodcut illustrations call

[10] Mott, *History of American Magazines*, 151.
[11] Bonner, "Hieronymus Pop and the Baby," 21.

Fig. 2. Illustration by A. B. Frost, "Hieronymus's Charge," *Harper's New Monthly Magazine* 61, no. 361 (June 1880), p. 20

attention to themselves as "subjective" and "contrived." Thus, I would argue that despite the stereotyped quality of the illustrations, they help call attention to the very fictiveness of the story itself, which is in the tradition of local color. This is particularly true of the illustrations of the mother, which are hugely stereotyped with minstrel-like large lips (fig. 2).

Local color was a genre of writing, mostly seen in the short stories that focused on rural regions of the United States, partly in response to what was beginning to be seen as a kind of cultural homogenization at the end of the nineteenth century. Local color stories, though, when they represented African Americans, were more than just nostalgia for what was passing away; they were political statements designed to show that African Americans were permanently incapable of joining the American polity, that they were to be forever relegated to the status of humorous not-quite-humans. My analysis of Frost's woodcuts, however, points to a level of sophistication on the part of readers and viewers, who would see the constructedness of the images and of the story, thus calling into question the verisimilitude of both.

Another example of this kind of illustration is from an essay about contemporary American caricature that appeared in *Scribner's Magazine* (fig. 3).[12] This illustration is by E. W. Kemble, who was also renowned in his time for his depictions of African American life. Since this illustration, "The Thompson Street Poker Club," is used in an article on caricature, it was understood in its time that these representations of African Americans were stereotypical exaggerations. Clearly, the exaggerated lips of all of the figures recall the lips of the mother in the previous illustration, and the overall depiction of happy gambling and cheating African Americans is an image right out of minstrel shows. To call attention to the constructedness of these images is not, however, to deny the power that they had to injure those who were being depicted. It is, on the other hand, to point to a certain sophistication on the part of the audience, which would realize that these were exaggerations, something that would not be the case, I am arguing, with images that were produced through the halftone process.

The Thompson Street Poker Club—"Whad yo' got?"
(From a drawing in *Life*, by E. W. Kemble.)

Fig. 3. Illustration by E. W. Kemble, "The Thompson Street Poker Club—'Whad yo' got?" *Scribner's Magazine* 6, no. 6 (December 1889), p. 736

[12] Mitchell, "Contemporary American Caricature."

The first halftone image I would like to consider is from an 1894 article in *Scribner's* by Octave Thanet, entitled "The Farmer in the South," from a series called "Sketches of American Types" (fig. 4).[13]

The General.

Fig. 4. Illustration by A. B. Frost, "The General," *Scribner's Magazine* 15, no. 4 (April 1894), p. 406

[13] Thanet, "Sketches of American Types."

This illustration is again by A. B. Frost, and the difference in quality between this image and the earlier ones is obvious. This halftone illustration has more of the quality of a photograph, of a transcription of reality, and it does not call attention to its own constructedness. Even though the illustrations of the Bonner story were highly and offensively stereotypical, I would argue that this image is even more damaging in its effect on the white audience. Clearly it is less obviously stereotypical—the old African American man is not a figure out of the long history of caricaturing African Americans—but the image is even more pernicious in its effort to reinforce and reinscribe what white Southerners and many Northerners thought was the only way that blacks and whites could occupy the same social space. While the figures are reverse images of each other (both men with canes in the right hands), the image encodes what was seen as the "proper" and "natural" relations between the races. The white man on the steps, from his position alone, is the superior, and the contrast of their clothes makes this hierarchy even clearer (the interesting reversal, though, is of age: the viewer would also accept as natural the subservience of the "old" African American to the "young" white man).

Even more problematic is another illustration from *Scribner's* of a scene from Thomas Nelson Page's novel, *Red Rock* (fig. 5).[14] Page was most well-known for his local color tales, stories that created a mythical, romanticized pre-bellum past in which kind and concerned masters took good care of contented slaves, and these freedmen then told the stories of their masters in the post-war era, expressing a proper nostalgia for slavery. As an old freedman says in Thomas Nelson Page's "Marse Chan" in what is a locus classicus of the celebration of slavery by the slaves who had been oppressed by it: "Dem wuz good ole times, marster—bes' Sam ever see! Dey wuz in fac'! Niggers didn' hed nothin' 't all to do."[15] Of course, the cultural work these tales did in the 1880s and 1890s was to help justify the political and economic repression of African Americans, who were believed to have degenerated under the stresses of the freedom they were thought to be constitutionally incapable of appreciating or employing. *Red Rock*, however, took his writing in another direction: the novel is a historical romance that begins in the pre-Civil War period, and that rewrites the history of the Reconstruction from the Southern point of view. It was "the most significant novel published by a Southerner in the 1890s, fifth on the bestseller list in the year of its publication,"[16] and it has been

[14] Thomas Nelson Page, *Red Rock: A Chronicle of Reconstruction*, chapters 17–21, *Scribner's Magazine* 23, no. 5 (May 1898): 613–32.
[15] Page, *In Ole Virginia*, 10.
[16] Bargainnier, "*Red Rock*: A Reappraisal," 44.

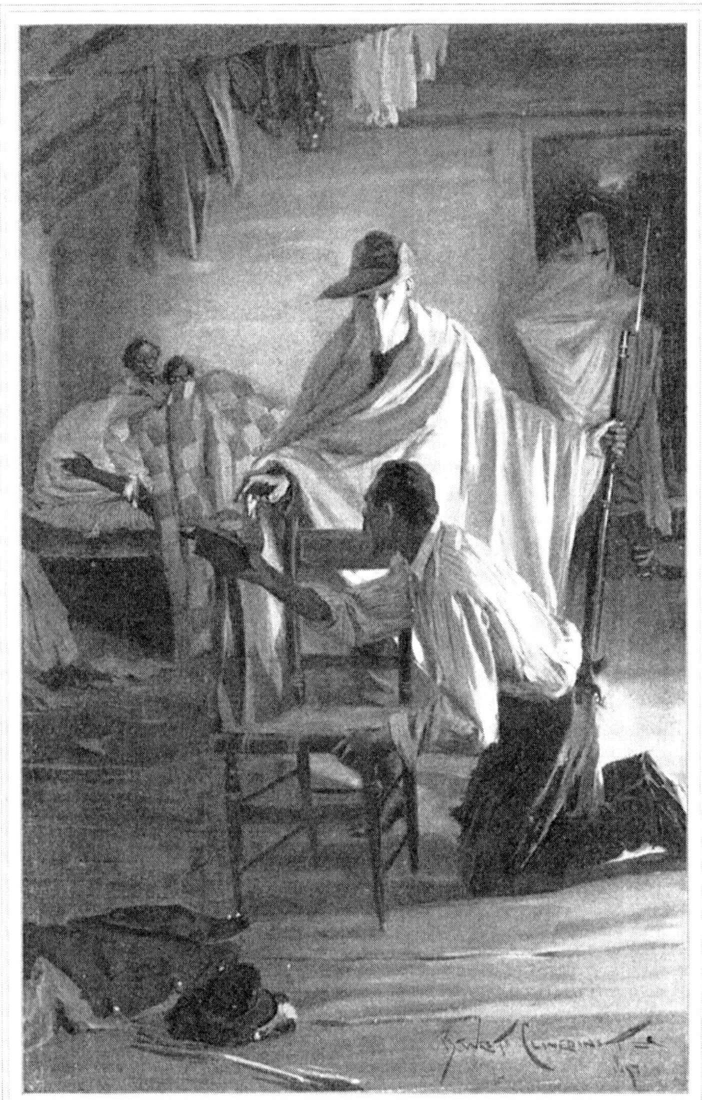

Ku Klux—"Awful forms wrapped like ghosts in winding sheets."—Page 628.

Fig. 5. Illustration by B. West Clinedinst, "Ku Klux—'Awful forms wrapped like ghosts in winding sheets,'" *Scribner's Magazine* 23, no. 5 (May 1898), p. 612

called the "clearest and most elaborate fictional description of Reconstruction from the Southern point of view."[17] By thus rewriting and revising Reconstruction, Page was trying to present Southerners as its victims, and was attempting to justify the terrorism of Southerners against Republicans and politically engaged African Americans.

For me, this illustration of two Ku Klux Klan members terrorizing an African American family has the opposite effect of what it would have had in its time: the viewer was intended to see this violence as a legitimate response to African Americans trying to impose themselves on white Americans, both in the social and political realms (notice the uniform of the African American man on the floor on the right next to him). The audience is not expected to feel sympathy for the family whose house is invaded by the terrorists, but to feel sympathy for the terrorists themselves. As the hero of the novel says: "We propose to obey the laws, but we do not mean to be governed by negroes, and if you attempt it you will commit a great mistake."[18] It would not be going too far, I believe, to connect an image like this to the Dunning school of Southern historiography, which argued that the violence against African Americans was justified by their misrule and their effort to dominate whites. Of course, this interpretation of the Reconstruction and its aftermath was a distortion of the historical record, as was this illustration of Page's novel. But that distortion, given the "reality effect" of the halftone image, was accepted as truth by its viewers. I would like to argue that the reality effect of halftone illustrations like these is one of the constituent elements that contributed to the explosion in the number of lynchings of African Americans in this period, peaking in 1892 with 156 lynchings throughout the United States, some 88 percent in the South. Joel Williamson has written that this "sudden and dramatic rise in the lynching of black men in and after 1889 stands out like some giant volcanic eruption on the landscape of Southern race relations."[19] The halftone image of African Americans and its literary brother, the local color Southern story and novel, worked in tandem to objectify African Americans and to justify their treatment as beings of lesser value.

For my last halftone image, I would like to explore the opposite side, the effort of African Americans to use this new technology to represent themselves, and, in looking at the effect of an image like this, I will be turning my argument so far on its head (fig. 6). This image comes from a 1906 issue of the *Colored American Magazine*, a race magazine with a very

[17] Gross, *Thomas Nelson Page*, 86.
[18] Page, *Red Rock: A Chronicle of Reconstruction*, 189.
[19] Williamson, *Crucible of Race*, 117.

From photo by Purdy, Boston.

THE YOUNG COLORED AMERICAN.

Fig. 6. Photograph by James E. Purdy, "The Young Colored American," *Colored American Magazine* 1, no. 5 (October 1900), frontispiece

small circulation that attempted to act as "an advocate for the black community."[20] In that sense, the *Colored American Magazine* was consciously offering itself as a counter-interpretation of the significance and position of African Americans, and it also tried to offer counter illustrations. And Sigrid Anderson Cordell has argued that this halftone image "attempts to align U.S. interests with those of the black community and reflects the magazine's aim to recover the role of African Americans in American history. The figure of the child evokes both a sense of optimism and an historical likeness to America's infancy."[21] If, on the contrary, this image had been seen by the majority of Americans, it would have been perceived as farcical and risible, as an assertion of and demonstration of the unsuitability of the young colored American to join the all-white polity. Rather than being, as Jussim argues, an enhancement of the real, rather than being "the illusion of seeing an actual scene, or receiving an objective record of such a scene," this halftone photograph would have been perceived by white viewers as comic fantasy. In the white audience's response to a photograph like this, the racial imaginary overpowers the reality effect of the halftone, so, as the image is co-opted into the racial imaginary, it turns into its opposite. This process can be seen at work most clearly in the reception of Charles W. Chesnutt's short stories that attempted to critique the tradition of plantation mythology. When Chesnutt tried, as I have written elsewhere, to alter the tradition of representation he had inherited, the works were turned, by the very racism they were critiquing, into support *for* racial ideology.[22] This insidious process has been elucidated by Slavoj Žižek when he wrote that "ideology really succeeds when even the facts which at first sight contradict it start to function as arguments in its favor."[23] Even though the editors of the *Colored American Magazine* intended this photograph as a contradiction of the dominant ideology, that ideology would have turned the meaning of the photograph on its head: what was an effort to "recover the role of African Americans in American history" would have been seen by a white audience as evidence for their being forever outside of that history. Therefore, whichever way African Americans were illustrated—whether they depicted themselves or whether they were represented by white illustrators—they could not escape from the trap of being perceived as lesser than, and ever subservient to, whites.

---

[20] Cordell, "'Case Was Very Black Against' Her," 59.
[21] Ibid., 52.
[22] Cf. Wilson, "Who Has the Right to Say," 18–35.
[23] Žižek, *Sublime Object of Ideology*, 49.

Works Cited

Bargainnier, Earl F. "*Red Rock*: A Reappraisal." *Southern Quarterly* 22 (1984): 44–53.

Bonner, Sherwood. "Hieronymus Pop and the Baby." *Harper's New Monthly Magazine* 61, no. 361 (June 1880): 20–25.

Brown, Michael. "The Popular Art of American Magazine Illustration, 1885–1917." *Journalism History* 98 (1998). http://ebscohost.com.

Cordell, Sigrid Anderson. "'The Case Was Very Black against' Her: Pauline Hopkins and the Politics of Racial Ambiguity at the *Colored American Magazine*." *American Periodicals* 16 (2006): 52–73.

Gross, Theodore. *Thomas Nelson Page*. New York: Twayne, 1967.

Harris, Neil. "Iconography and Intellectual History: The Halftone Effect." In *Cultural Excursions: Marketing Appetites and Cultural Tastes in Modern America*, 304–17. Chicago: University of Chicago Press, 1990.

Jussim, Estelle. *Visual Communication and the Graphic Arts: Photographic Technologies in the Nineteenth Century*. New Providence, NJ: Bowker, 1983.

Mitchell, J. A. "Contemporary American Caricature." *Scribner's Magazine* 6, no. 6 (December 1889): 728–45.

Mott, Frank Luther. *A History of American Magazines, 1885–1905*. Cambridge, MA: Harvard University Press, 1957.

Page, Thomas Nelson. *In Ole Virginia, or Marse Chan, and Other Stories*. Ridgewood, NJ: Gregg, 1968.

—. *Red Rock: A Chronicle of Reconstruction*. Ridgewood, NJ: Gregg, 1967.

Thanet, Octave. "Sketches of American Types: The Farmer in the South." *Scribner's Magazine* 15, no. 4 (April 1894): 399–410.

Williamson, Joel. *The Crucible of Race: Black and White Relations in the American South since Emancipation*. New York: Oxford University Press, 1984.

Wilson, Matthew. "Who Has the Right to Say: Charles W. Chesnutt, Whiteness, and the Public Sphere." *College Literature* 26 (1999): 18–35.

Žižek, Slavoj. *The Sublime Object of Ideology*. London: Verso, 1989.

TECHNOLOGY FOR THEATER: HOW INVENTIONS CHANGED 19TH-CENTURY AMERICAN THEATER

RADOSŁAW RYBKOWSKI

The history of American professional theater started for good just after the American Revolution was over. Companies of players were reestablished and reorganized, new theater buildings were constructed. However, it was the 19th century that witnessed a radical and unprecedented development of this form of artistic expression. That was partially due to the growth of the cities, the changes in the means of transportation and communication (railroads and the telegraph) and simply the demographic growth of the US population: the more inhabitants there were, the more potential theatergoers could be attracted.[1]

Growing demand for theatrical entertainment of some quality forced the changes in the ways it was advertised and presented to the public. People in the United States no longer agreed to be seated on wooden benches, in a rather dark auditorium. American economic stability and prosperity influenced the habits of 19th century theater lovers; they wanted not only to listen to dramatic performances (although that is the origin of the word "audience": at first, people in the theater were supposed to listen, *audio* in Latin), but they also wanted to see the performances and to see them clearly.[2]

This demand of the audience does not seem exaggerated. Many theater critics, historians and artists of the theater say that the most unique feature of this form of art is a two-way artistic communication. Unlike any other form of art—this was particularly true in the 19th century—theater demands the presence of a live performer and a live audience. In theater a performer or performers must communicate, sometimes using not only words but gestures or mimics instead; the audience must listen and watch, must follow the plot of the play, and the reactions of the audience provide

[1] Wilson, *Three Hundred Years*, 75–78; Brockett, *History of the Theater*, 476.
[2] McConachie, *Melodramatic Formations*, 46–75.

actual feedback to the artists. Memoirs of actors and actresses prove that silence of the enchanted audience or laughs of people having fun are not only the measure of success, but also a necessary part of theatrical communication. Otherwise, there is no theater at all.[3]

Since the end of the 18th century, political stability and economic prosperity in the newly established United States gave the people enough time, energy and money to attend performances. Thanks to these factors, theater buildings—both the auditorium and the stage—began to expand in size. The enlargement of the stage resulted in one problem, however. The quality of the theater building of that time was measured neither by seat capacity nor by the width of the stage. It was the depth that really made the theater impressive: sufficient depth helped to arrange an interesting set design and to present eye-catching choreography. Growing depth, on the other hand, created new challenges.[4]

One of the best examples of the complexity of the problem was Chestnut Street Theater in Philadelphia. The theater was built in 1793 and closed almost at once because of the yellow fever epidemic; the regular season started the following year. The theater could seat ca. 1,200 spectators and had a stage 11 meters wide and 21 meters deep, which made the theater quite large, even compared to European ones. At the end of the 18th century the most common way of illuminating was by candlelight or by oil lamps. The intensity of the light, as is easy to imagine, was insufficient; the backstage of the Chestnut Street Theater, for example, could not be used properly, since barely anything could be seen. Therefore, it was a common practice of actors and actresses of that time (until the second half of the 19th century) to come to the proscenium (the part of the stage closest to the audience) just to be properly lit and be visible to the spectators.[5]

The situation changed in 1816, just in the Chestnut Street Theater, although meanwhile some other large theaters in the USA were built, including the prominent Park Theater in New York City in 1806. The managers of the Chestnut Street decided to experiment with a new invention: the gas lamp. At first it was used exactly the same way as candlelight. Because it was not so easy to light up and then blow out the candles, the buildings—both the stage and the auditorium— remained illuminated during the whole performance. Until the early 1820s gas lamps

---

[3] Styan, *Drama, Stage and Audience*; Dudden, *Women in the American Theater*, 14–16.

[4] Brockett, *History of the Theater*, 369–72.

[5] Henderson, "Scenography, Stagecraft, and Architecture in the American Theater: Beginnings to 1870," 392–94; Brockett, *History of the Theater*, 473.

remained switched on during the whole evening. People complained that due to the lamps it was unbearably hot, and the smell of smoked gas was also distracting.[6]

New opportunities provided by the use of gas light were very soon discovered. It was no problem to switch off the light above the auditorium during the performance and to switch it on again during entr'actes. Since gas gave much stronger light than candles, for the first time the whole stage, including far backstage, could be used for theatrical productions.[7]

Illuminating the backstage was another factor that resulted in switching off the auditorium light. If the person sits in the dark, it is a natural reaction to focus on illuminated space. Thus, these two changes—better lighting condition onstage (the whole stage) and switching off the light over the auditorium—eventually changed the patterns of theatrical artistic communication. Previously, the audience could only hear the performers and to a great extent only imagine the gestures, costumes and stage-setting; now, the theatergoers were able to actually see what was happening on the stage. Treated as a kind of novelty, elaborate set design became one of the major theater attractions in the 1820s and 1830s.[8]

It was not Chestnut Street Theater, however, that was the best example of this new tendency in theater development. In 1826 the new Bowery Theater was built in New York City as competition for Park Theater, which was quite prosperous under the management of Thomas A. Cooper and Stephen Price. The Bowery Theater was much bigger, its capacity was more than 3,000 seats; after the fire in 1828 it was rebuilt and its capacity reached almost 4,000. It is quite understandable that in such a building, without artificial support of microphones and speakers, it was not the voice of the performer that could attract the attention of the audience.[9]

The Bowery Theater (especially in the years 1830–1848, during the times of Thomas S. Hamblin as a manager) focused on visual effects and elaborate stage-setting as the main attraction. In the advertisements published in the New York City press of that time new settings seemed to be more important than the quality of plays or the talent of the performing actors and actresses.[10] Very soon, New York City audiences were more attracted to performances which offered striking visual effects than to those featuring an intelligent plot. Especially the less educated working class of that time (a typical audience of the Bowery Theater) was very

---

[6] Odell, *Annals of the New York Stage*, vol. 2, *1798–1821*, 485.

[7] James, *Old Drury of Philadelphia*, 25–27.

[8] Hodge, *Innovations in Stage*; Oenslanger, *Stage Design*, 169–70.

[9] Morris, *Curtain Time*, 207–9.

[10] Odell, *Annals of the New York Stage*, vol. 3, *1821–1838*, 509–11.

eager to accept this type of theatrical entertainment. Therefore, the stage of the Bowery was crowded with plays full of battles, gunshots, volcano eruptions, earthquakes and other disasters that pleased the theatergoers. For the purpose of such performances special side gates, used by artillery and cavalry, were built. For many years the Bowery was nicknamed "Bowery slaughterhouse" due to the style of the presented productions.[11]

The performance of the early Republic's favorite play, *Bunker Hill*, is a perfect illustration of the quality of theatrical productions that could be achieved partially thanks to the use of gas light, since without proper illumination the noise of the battle could be heard, but spectators could not enjoy the panoramic view of the revolutionary struggle, here described by Bruce McConachie:

> The English march in two divisions from the wings to the foot of the hill. At this point American and English soldiers fired rifles at each other for several minutes . . . . As one English unit retreated and another took up battle, the playwright reminded the manager to open all windows near the stage to let out the smoke. The Americans finally left their shelter and met the British mounting the hill, which gave room for effect, if the scuffle be nicely managed. While two or three Englishmen were rolling down hill, a backdrop having some houses and a meeting house painted in fire, with flame and smoke issuing from it, should be raised. Small cannon were to be frequently fired during this battle scene which lasted for twelve or fifteen minutes.[12]

The selection of the plays and the new approach to stage-setting were clear reflections of the theater's growing popularity. Since the 1820s, theater became a popular entertainment of the American society, and the only real mass entertainment as well. The invention of gas light and its use by the theater resulted not only in changes considered by American elites as the deterioration of theatrical culture. With better light, it was no longer possible to use shabby, worn-out costumes. In some cities, managers had to adopt some regulations since actors' costumes sometimes happened to have holes in less respectable places, a fact previously unnoticed.[13]

The use of gas light had yet another side-effect. Until the early 1820s theaters were a typical place where American prostitutes could find clients. A bar, serving alcoholic beverages, was usually located in the third tier of the gallery and this was an ideal place for the prostitution of that time. The growing size of the theater buildings demanded the influx of new audiences: people that were previously uninterested in or even afraid

---

[11] Grimstead, *Melodrama Unveiled*, 54–56.
[12] McConachie, *Melodramatic Formations*, 77.
[13] Grimstead, *Melodrama Unveiled*, 91–93.

of theatrical entertainment. Willing to increase profits, theaters started to present performances not three times a week, as it had been in the 18th century, but six times a week; therefore, the presence of any offensive and notorious group among the audience could no longer be tolerated. At first, the use of better light and switching off the auditorium light was considered to strengthen the "third tier culture" even further, but actually it had an exactly opposite effect. In the dark, prostitutes could not look for clients and the potential clients could not find mistresses. The light of the stage attracted even this group's attention, and thus the "third tier culture" perished.[14]

The technology used in theater had a deep impact on the way the plays were performed and perceived by spectators. New inventions in lighting changed the pattern of theatrical communication and since 1816 the visual effects, gestures, mimics, scenography, and choreography started to dominate the perception of the theater. This was further changed in the 1840s by two new inventions. One of them was the so-called gas table. It was a type of switchboard that enabled one to control the light in the whole building (including the stage and the auditorium) from one place. This allowed control of gas supply to separate parts of the stage and, in consequence, made entirely new effects possible. The other was the lime lamp, which produced an extremely powerful light beam and was used as the first spot light in theater. Thanks to the lime lamp, not only the whole stage but also the face of the actor or actress could be illuminated. That helped to focus the attention of the audience even more. The use of gas light became much easier in the 1840s, not only because of these new inventions. During this time, cities started to build municipal gas pipe networks, so it was easier to find a secure supply of gas at no extra cost.[15]

Thomas Alva Edison's new invention—the electric light—was not as important for the development of theatrical art. Although it made the use of light even easier than in the case of the gas light, the core of the technique used in theater remained the same: adjustable light, controlled by one central switchboard and the spot light used to illuminate performers or a particular part of the stage-setting. The only important difference was connected to the side-effect of the gas lamp: with electric light there was no extra heat or smell. On the other hand, the electric light was not as reliable at first. Quite simply, light bulbs would often blow out during performances. Electric light was first used in American theater in 1882, in Boston Bijou Theater; the real importance of electric light was started by

[14] Dudden, *Women in the American Theater*, 78.
[15] Leverton, *Production*, 20–27.

the Kliegel brothers. They established a company specializing in theatrical illuminations and the success of the Metropolitan Opera House, where they installed a lighting system with 96 resistance dimmers (and 20 additional dimmers for the house lights), initiated the triumphant pace of the electric light.[16]

Other inventions of the second half of the 19th century are closely connected with one man: Steele MacKaye. He was one of the most versatile figures in the history of the American theater—an actor, director, theater manager, painter, set designer, teacher, playwright, architect, and inventor. The influence of MacKaye on the development of the American theater could have been much greater if only he had been more patient in realizing his ideas.[17]

In 1879, Steele MacKaye became the manager of the Winter Garden Theater in New York City. The building was partially based on his own plans and he was the originator of some inventions used there for the first time in the history of theater. He created the hydraulic stage elevator—a very sophisticated piece of machinery that enabled fast changes of the stage-setting. Until 1879, there was a limited number of complete changes of the setting, usually three per act at the maximum. If more changes were necessary, the performance had to have an entr'acte. When the curtain was dropped and the audience left for the lobby, the men onstage could change the setting and prepare new sets of decorations. That practice influenced playwrights; they usually restricted the number of locations to just three in one act, so the play could be presented onstage without problems. This can be easily noticed in dramas of the Romantic period.[18]

MacKaye's stage elevator was a two-storey mechanical device. During the performance, one of the stories was used as the actual stage, with complete stage-setting. During the performance, however, the technicians could prepare the setting for the next scene on the other storey. Whenever a change of setting was necessary, the whole elevator moved down (or up, depending on the initial position of the elevator) at an incredible speed: in less than one minute. The public's fascination with 55-second scene changes of totally realistic three-dimensional scenery was sufficient to make the Winter Garden Theater a success.[19]

Steele MacKaye's invention, used in other theaters as well, changed the attitude of American dramatists. They were no longer restricted by theatrical technology; they could use their imagination freely and the

[16] Henderson, "Scenography, Stagecraft, and Architecture," 508–11.
[17] Wilson, *History of American Acting*, 99–103.
[18] Grimstead, *Melodrama Unveiled*, 120–36.
[19] MacKaye, *Epoch*, 323–26.

locations portrayed in the plays could be as numerous as the artistic vision required. Previous limitations had forced authors to balance the length of the scenes in one act, so that one location was not shown for too long and another for too short a time. The stage elevator made this balancing redundant and the author's talent became the only restriction in American drama.[20]

MacKaye's innovations helped not only to reduce the limitations in staging dramas. He was also deeply concerned with the quality of the audience's experience. The Chestnut Street Theater, built in 1794, had only wooden benches, and some of the spectators even stood during performances. Since mid-19th century, theaters needed to provide some comfort for the audience. After the auditorium light was switched off, spectators were expected to sit and to watch the performance, not to talk to the neighbors, walk or order food or drinks too loudly. Theaters had to arrange comfortable seats for the public; however, more comfortable seats had a negative impact on the capacity of theater buildings. MacKaye created something brilliant, still in use in theaters and movie-theaters worldwide: the folding chair. Thanks to this invention, two aims could be achieved at once: an increase in comfort and in the number of spectators. Therefore, it is not surprising that folding chairs very soon became one of the most typical features of all American theaters.[21]

The other invention of Steele MacKaye was also connected with the comfort of spectators and it reflected the style of the Victorian era. In the second half of the 19th century, women's millinery could be of enormous size. If a woman wearing such a hat came to the theater, people in the rows behind her could barely see anything on the stage. To overcome this inconvenience, MacKaye created hat hallstands and used them for the first time in the Winter Garden Theater. Thus, female audience could bring attractive hair dresses but they no longer distracted others from enjoying the show.[22]

19th century theatrical innovations had impact on both the stage and the audience. Improved technology radically reduced the limitations in stagecraft practice, which enabled spectators to fully admire the stage-setting and plot without being unnecessarily disturbed. It is definitely not true that technology changed the core of theatrical experience: the magic of communication between the actor/actress and the audience remained unchanged. Technology, however, had influenced the patterns of communication.

---

[20] Ibid., 290–300.
[21] Henderson, "Scenography, Stagecraft, and Architecture," 487–513.
[22] MacKaye, *Epoch*, 298–310.

Since the 19th century, thanks to the new lighting conditions, visual effects became much more important and the spectators could become real spectators, that is those who actually watch. The audience sat down and followed the plot, no longer using mainly their ears, but their eyes. This eventually led to new patterns of audience behavior; interested in the action onstage, spectators were no longer so interested in other attractions (such as prostitutes and the "third tier" culture). The stage elevator constructed by Steele MacKaye made the visual appearance of the performance even more important and easier to present on the stage.

New theatrical inventions started as novelties but very soon were widely accepted and helped to develop further the dramatic and theatrical art in the United States as well as in other countries. Therefore, examining the technology could be a fascinating adventure since it helps us discover what continues to be really important in theater today.

Works Cited

Brockett, Oscar G. *History of the Theater*. Boston: Allyn and Bacon, 1982.

Dudden, Faye E. *Women in the American Theater, Actresses and Audiences, 1790–1870*. New Haven, CT: Yale University Press, 1994.

Grimstead, David. *Melodrama Unveiled: American Theater and Culture, 1800–1850*. Chicago: University of Chicago Press, 1968.

Henderson, Mary C. "Scenography, Stagecraft, and Architecture in the American Theater: Beginnings to 1870." In *The Cambridge History of American Theater*, edited by Don B. Wilmeth and Christopher Bigsby, 1:373–423. Cambridge: Cambridge University Press, 1998.

—. "Scenography, Stagecraft, and Architecture." In *The Cambridge History of American Theater*, edited by Don B. Wilmeth and Christopher Bigsby, 2:487–513. Cambridge: Cambridge University Press, 1998.

Hodge, Francis. *Innovations in Stage and Theater Design*. New York: American Society for Theater Research, 1972.

James, Reese D. *Old Drury of Philadelphia: A History of the Philadelphia Stage, 1800–1835; Including the Diary or Daily Account Book of William Burke Wood, Co-manager with William Warren of the Chestnut Street Theater, Familiarly Known as Old Drury*. Philadelphia: University of Pennsylvania Press: 1932.

Leverton, Garrett H. *The Production of Later Nineteenth Century American Drama: A Basis for Teaching*. New York City: Teachers College, Columbia University, 1936.

MacKaye, Percy. *Epoch: The Life of Steele MacKaye, Genius of the Theater, in Relation to His Times and Contemporaries; A Memoir by His Son*. New York: Boni and Liveright, 1927.

McConachie, Bruce A. *Melodramatic Formations: American Theater and Society, 1820–1870*. Iowa City: University of Iowa Press, 1992.

Morris, Lloyd. *Curtain Time: The Story of the American Theater*. New York: Random House, 1953.

Odell, George C. D. *Annals of the New York Stage*. Vols. 15. New York: Columbia University Press, 1927–49.

Oenslanger, Donald. *Stage Design: Four Centuries of Scenic Invention*. New York: Viking Press, 1975.

Styan, John L. *Drama, Stage and Audience*. Cambridge: Cambridge University Press, 1975.

Wilson, Garf B. *Three Hundred Years of American Drama and Theater: From Ye Bear and Ye Cubb to Chorus Line*. Englewood Cliffs, NJ: Prentice-Hall, 1982.

"INCOHERENT NARRATIVE, JAGGED TECHNIQUE, DISTURBING INTENSITY OF EMOTIONS…": THE CASE OF JOHN CASSAVETES

ELŻBIETA DURYS

In the English-speaking world, the theory of the *auteur* and authorship in cinema emerged and developed concurrently with the work of John Cassavetes (1929-1989). Therefore, it seems paradoxical that someone who can certainly be regarded as an author, in the colloquial and most obvious meaning of the word—who wrote his own scripts, directed, held the camera, edited the material, starred in his films and often promoted and distributed them himself—should not only be omitted from the examples provided by academic literature devoted to the subject, but also be considered unworthy of a serious critical study until the 1980s.[1] Nowadays, Cassavetes is treated as one of the luminaries, perhaps even the initiator of American independent cinema in its current understanding, while his debut (*Shadows*, 1959) is often held up as a shining example.[2] The initial lack of interest on the part of American critics and audiences can be variously explained. Ray Carney emphasizes the limited availability of Cassavetes's films, as well as their novelty.

In my opinion, the main reason lay in the American viewers' attitude, shaped by the Hollywood film practice which was dominant at the time. The audiences were accustomed—as far as so-called "serious" cinema was concerned—to three things: firstly, it must deal with Important Subjects; secondly, those matters must apply to a larger portion of the society (hence

[1] 1985 saw the publication of the first monograph, authored by Raymond Carney, a Boston University professor; cf. Carney, *American Dreaming*. In later years Carney returned to the subject several times. Among his publications are *The Films of John Cassavetes*, and a bulky volume of Cassavetes's own statements (*Cassavetes on Cassavetes*).

[2] Cf. Fine, *Accidental Genius*; Tzioumakis, *American Independent Cinema*, especially pp. 174–77; also King, *American Independent Cinema*, 71.

the tendency to generalize); finally, the ending should bring a resolution of the problem. Those three elements were usually combined, thus substantiating the consequent accusations of didacticism, simplification or even reproducing the dominant ideology, which were leveled at the productions of—the name speaks for itself—the Dream Factory in the 1970s. Cassavetes was consistent in ignoring such an approach to reality. This, however, is not to say that the matters which he addressed were trivial or that the problems discussed by him only concerned a small fragment of the society. At the heart of his interests were common people, such as siblings, husbands, wives, children or friends, affected by common matters such as puberty, racial issues (*Shadows*), marital conflicts and adultery (*Faces*), the situation of women (*Minnie and Moskowitz, A Woman Under the Influence*) or aging and mortality (*Husbands*, *Opening Night*). In one of his interviews, the director stated: "I really don't try to communicate about so-called 'important subjects', I talk about the *people* involved in those subjects."[3] However, Cassavetes never generalized. Each situation—even as mundane as a shared breakfast—was treated as unique and special in his films. This is the source of the intensity which marks those works and which may sometimes be difficult to endure. Most importantly, however, he did not offer ready solutions.

Arguably, the initial rejection of Cassavetes's *oeuvre* was connected, first and foremost, with his unconventional use of the medium. This American director of Greek origins employed film technique in a manner unheard of in American feature productions at the time. He emphatically rejected the zero degree style, developed by Hollywood and regarded as dominant. He also distanced himself from art cinema, both the New York avant-garde variety and the European pictures imported from overseas. Cassavetes's innovation in the use of film technique can be narrowed down to three areas: firstly, he "borrowed" the style of the documentary from the American direct cinema; secondly, he constructed scenes (understood as the film's dynamic units) in a way which differed considerably from the previous practice; lastly, his employment of genre formulas was truly singular.

As a trend in documentary filmmaking, direct cinema dates back to the 1950s. Mirosław Przylipiak claims that the shift of emphasis—the replacement of the dominant, propagandist-persuasive model by observational tendencies—was caused by a convergence of two factors. One was the technical improvement of the film equipment. The increase of tape sensitivity, the introduction of lightweight cameras and of portable

---

[3] Carney, *Cassavetes on Cassavetes*, 302.

sound equipment which made synchronous recording of image and sound possible not only allowed physical movement, but also—more importantly—a more spontaneous following of events. Another factor contributing to this change was the particular sociopolitical atmosphere. The new form of audiovisual presentation, according to Przylipiak, was the result of "the political model of liberal democracy . . . with its characteristic ethos of independent opinion and tuning into the voice of the nation . . . ."[4] As a trend, direct cinema became the embodiment of such tendencies in the US.

An in-depth analysis of the filmmakers' own statements and their works allowed Przylipiak to reconstruct the poetics of the trend, which was based on five elements. At its heart lay the injunction to record events the way they appeared and to restrict the filmmaker's action to "being on the site of events."[5] The goal of the Drew Associates[6] was to capture the truth of the occurrence. This objective was supposedly abetted by—and this was the second element—limiting the crew to two or three members and working without any preliminary assumptions (the deadliest enemy was the script, which was seen as imposing not only the content, but also the distribution of emphasis and meaning). Thirdly, the filmmakers were hostile towards various forms of interfering with reality, such as staging, reconstruction, provocation or interviews, in which—they maintained—it was possible to manipulate the person and their behavior through the questions. Fourthly, they claimed that the relation between the filmmaker and the filmed person ought to be based on openness, respect and sincerity. Finally, they believed that observation in itself allowed reality to divulge the truth about the events and the participants. This, in turn, had its consequences on the level of editing: the amassed material was supposed to reveal its inherent form of its own accord.[7]

Cassavetes took the poetics of direct cinema and adjusted it creatively to fit the feature film.[8] Undoubtedly, this was the starting point in the

[4] Przylipiak, *Poetyka kina dokumentalnego*, 131.
[5] Ibid., 137–8.
[6] The Drew Associates was a collective formed by documentary filmmakers who used methods of direct cinema in their works.
[7] Ibid., 137–51; cf. also Przylipiak, "Kino bezpośrednie," pts. 1 and 2; Przylipiak, *Kino bezpośrednie, 1960–1963*.
[8] This is also the opinion of Marek Hendrykowski (cf. item "direct cinema" in Hendrykowski, *Słownik terminów filmowych*, 34) and George Kouvaros, *Where Does It Happen?* Geoff King also writes about the influence which the documentaries of the time had on independent cinema (among others, on

making of *Shadows*; and yet, the crux of the matter is that one runs into difficulties when attempting to prove it definitely. Cassavetes himself never addressed this issue. He did not speak at length of his inspirations; he was even irritated by such inquiries. Furthermore, the declared poetics of direct cinema was far removed—as indicated in Przylipiak's analyses—from the actual productions. Hence the dilemma as to what should be regarded as the point of reference: the filmmakers' declarations or the completed works? Lastly, Cassavetes's own poetics evolved as well. It is enough to consider the huge difference between his first *auteur* film, the aforementioned *Shadows* (1959), and the follow-up, *Faces* (1968); understandably, the gap widened in his subsequent works.

Having at his disposal the Hollywood productions of the era, of which he spoke with unreserved contempt,[9] and, at the other end of the spectrum, the experiments of the New York avant-garde, Cassavetes chose his own path when making *Shadows*. From a temporal perspective, this path could be said to resemble the aesthetics of direct cinema the most. This tendency manifested itself on the technical plane and, as a result, on the level of style. Location shooting, filming in the streets and public spaces of New York, restricting the crew to two or three members, using private apartments for the purposes of the work: all this resulted from a prior choice at the heart of which lay—as in the case of the proponents of direct cinema—the desire to capture the truth of events. While the latter achieved their goal by "entering the world holding a camera," Cassavetes, as an author of fiction films, could not simply point the lens at a selected fragment of reality; he was forced to arrange this reality in a particular way. The truth which he wished to convey in his debut was one concerning the initiation into adulthood: the collision of dreams and expectations with harsh reality. The issue was complicated further by the fact that the main characters—the siblings Hugh, Ben and Leila—were Mulattos. This is why the work is regarded as one of those which reveal, with astonishing finesse, "just a little problem . . . with the races, that's all," as the elder brother puts it in a scene from the film.

In the case of *Shadows*, the lack of preliminary assumptions resulted not so much in abandoning the script and shooting improvised scenes[10] as

---

Cassavetes's productions), but in doing so, he refers to the French variety, i.e. *cinema verité* (King, *American Independent Cinema*, 121–22).

[9] He even launched a direct attack on Hollywood in an article published in *Film Culture* (Cassavetes, "What's Wrong with Hollywood," 22–23).

[10] The myth of improvisation as creative method was abetted by Cassavetes himself, who inserted a card which read "The film you have seen was an improvisation" at the end of the work. I have previously discussed the issue of

in an openness characterizing particular scenes as well as the entire work. The film ends at a moment when the viewers have the most questions and are unable to find answers to any of them. As mentioned previously, the absence of ready solutions was one of the strategies employed by the author. Such was his notion of capturing the truth of events: life never ossifies into formulas, but flows and evolves constantly. The picture allowed one to stop for a moment and examine a particular problem in the form which it assumed in an individual existence—that was the unique quality in the American's approach to filmmaking. This may explain why Krzysztof Mętrak, shocked by this painful individualism which meant rejecting all forms of expression, called the reality of the characters "Cassavetes's pulp."[11]

The champions of direct cinema preferred long takes, in an attempt to eliminate editing, which was regarded as a means of creating meaning, or even, as some believed, of manipulating reality. However, extending the length of the take—Przylipiak argues—was often merely declared and not actually followed in practice: "The length of the take did not exceed the average typical of films of the period. For instance, *On the Pole* (1960), at its length of 53 minutes, contains 369 takes. This means that the average take length is 8.6 seconds, i.e. slightly beneath the length in a typical American feature film of the era, which was 9.3 seconds."[12] The case of Cassavetes is similar: the editing of *Shadows* involved a standard number of cuts and even some unusual camera setups (especially in the party scenes at the siblings' house). This would change in the films to come. The takes would be prolonged considerably, while the camera setup would remain unchanged throughout the entire scene, often lasting more than ten minutes.

Another factor which contributed, on the technical plane, to the misunderstanding and rejection of Cassavetes's work by the American audiences, was the construction of the scene which was different from standard practice. The principal rule which emerged during the director's compulsory period of passivity and before the making of the second entirely *auteur*-driven film, *Faces*,[13] was to focus not so much on the truth

improvisation in the director's output in an article entitled "Improwizacja w *Cieniach* Johna Cassavetesa."

[11] Mętrak, "Miazga Cassavetesa," 62–63.

[12] Przylipiak, *Poetyka kina dokumentalnego*, 146.

[13] Cassavetes's filmography contains two other pictures which are often credited to him and inserted between *Shadows* and *Faces*, namely *Too Late Blues* and *A Child Is Waiting*. The director worked on those films in Hollywood, where he had been

of events as the truth of feelings and emotions which accompanied the events.[14] Cassavetes believed that he would achieve this if he adopted a different approach to scene shooting. He regarded the actor as central. Ben Gazzara, a longtime friend and a well-known Hollywood personality, maintained that the reason for this may have been Cassavetes's own experience in the profession and his frequently voiced dissatisfaction with the treatment of actors in Hollywood (and elsewhere).[15] In his own productions, the actor was a subject to whom everything else on the set—the camera setup, the lighting, the length of the take, the time and space—was subordinated. The scenes were shot on location. Often he would rent a house for this purpose or made use of actual apartments. For instance, *A Woman Under the Influence* was made in a house in Los Angeles, rented specially for the filming needs. This allowed the people involved to start feeling comfortable in a particular place, to get accustomed to it, to familiarize themselves with it and develop a sense of home. The actors' comfort was also heightened by abandoning the three-point lighting system. Whereas in Hollywood productions the movements of the actors were subordinated to the lighting setup, Cassavetes had the set thoroughly lit, so that nothing restricted the actors' behavior, their movements and gestures. Moreover, the actor did not have to play to the camera; thanks to the singular shooting method, he or she could move about freely.

Whenever the financial situation made it possible, Cassavetes would employ two cameras simultaneously. One was immobilized, fixed in a given position so as not to obtrude, while another, light and portable, was handled by the cameraman (most often Cassavetes himself) in order to—as the artist himself put it—"pick up details." In this way close-ups were achieved. The main camera was frequently equipped with a zoom lens, so that it could, if necessary, approach the actor at a moment of emotional tension without physically encroaching on his or her intimate space. This is particularly visible in the scene from *Faces*, where the Frosts, already in bed, perform their ritual of laughter, jokes and caresses. The viewer is treated to extreme close-ups of body and face fragments, and finally Maria's deeply moving look as she slips into solitude. Often, however, Cassavetes would try something entirely different. Taking advantage of the fact that he was holding the other camera, he would approach the actor, and, if he or she failed to react as expected, the director would whisper

summoned after the artistic success of his debut. Nevertheless, he was eventually barred from both on account of breaking the Dream Factory rules.

[14] Cf. Carney, *American Dreaming*, 88.

[15] His quarrels with Roman Polański on the set of *Rosemary's Baby* have become the stuff of cinematic legend.

something in the performer's ear, thus mobilizing them to be more active. Many complained about this unusual method; and yet—as they were quick to admit—the final effect would exceed their wildest expectations.[16]

The freshness, the novelty as well as the rejection of hackneyed patterns and schemata was Cassavetes's priority. The negative point of reference was invariably Hollywood with its attendant style. However, despite their declarations, there existed aspects of traditional cinema from which neither he nor the Drew Associates were able to liberate themselves. The latter—Przylipiak emphasizes—could not eradicate elements of romance or the structure of conflict and strife from their productions;[17] Cassavetes was plagued by the same difficulty. In his films, as Ray Carney expertly demonstrates, Cassavetes uses the heritage of genre films.[18] At times his references are explicit, e.g. to gangster film in *Gloria* (1980) or to *film noir* in *The Killing of the Chinese Bookie* (1975). He is more subtle in his use of romance and melodrama (*Faces*, 1968; *A Woman Under the Influence*, 1974) or screwball comedy (*Minnie and Moskowitz*, 1971). It is difficult to determine to what extent this practice is intentional and to what extent it results from being raised in a particular culture. Nonetheless, even here Cassavetes is never simply an imitator; his use of existing formulas is invariably creative.

As previously stated, the so-called "serious" Hollywood cinema was governed by three rules: firstly, it had to deal with Important Subjects; secondly, those subjects had to apply to a larger part of the society; finally, the ending was supposed to bring a resolution of the conflict, in a form which was readily applicable to the viewers' own lives. The tendency was most obviously visible in genre films; this led to accusations, leveled by ideological criticism, to the effect that such motion pictures reproduced the dominant ideology and assured the viewers that no other solution save the one offered by the system and genre films was possible. Cassavetes managed to undermine this way of thinking by inscribing the formulas with content which exploded them from the inside. Let us invoke three films as examples. While *Gloria*, made towards the end of the 70s, alludes to gangster films on the level of theme, iconography and characters, the depiction of gang warfare allows Cassavetes to address the issue of maternity and its place in a woman's life. An aging mob girl decides to forsake the privacy of her home and her quiet life which she has earned by

---

[16] Cf. *Cassavetes on Cassavetes*, 287–88; Carney, *American Dreaming*, 102, 188; Carney, *Films of John Cassavetes*, 137, 226.

[17] Przylipiak, *Poetyka kina dokumentalnego*, 150.

[18] Cf. Carney, *American Dreaming*, 1, 142–50, 226, 276; Carney, *Films of John Cassavetes*, 215–16; King, *American Independent Cinema*, 177.

saving money in order to help a Puerto Rican boy whose family was killed after his father embezzled mafia money.[19] The pair's getaway, as well as the ensuing chase, become a pretext for a tale about the choices which people make, the price which they pay and the ways in which opening up to another (in this case, a child) may change one's life. One more case in point is *The Killing of the Chinese Bookie*: by employing the aesthetics of *film noir*, Cassavetes tells a story of maturity—or rather an escape from it—in the life of a man in his forties. *Minnie and Moskowitz*, ignored both by the audiences and the critics upon its release, is, on the surface, a story about the encounter and love between two highly incompatible people. Minnie is educated, intelligent, attractive and lonely. Moskowitz, on the other hand, is a freeloader, who earns money parking the cars belonging to the customers of a restaurant and lives off of his mother. Despite their differences, the two fall in love and—in accordance with the rules of courtship screwball comedy—they start a family. Analyzed today, the film reveals two clichés, firmly rooted in the society and confirmed by the solutions offered by the genre. First of all, the only thing which is able to lend sense and meaning to a woman's life is the love of a man; furthermore, to keep this love, a woman will compromise, rejecting the values which she holds dear. Thus, the use of genre formulas allows Cassavetes to lay bare the thought patterns which control the collective conscience.

Writing about *auteurs* in American cinema of the 1970s, David Cook included Cassavetes, along with Terrence Malick, in the category of "eccentrics."[20] This decision was not really motivated by the fact that the director remained outside the Hollywood system (after the traumatic experiences of *Too Late Blues* and *A Child Is Waiting* he rejected all proposals which originated in the Dream Factory). His distinction and uniqueness are attested to not only by "a disturbing intensity of emotion that is virtually unique in the American cinema,"[21] but rather, first and foremost, by a consistently innovative use of the film medium in its three aspects, discussed in this paper. The fact that Cassavetes refreshed, perhaps even revolutionized the use of film technique was due not only to his adaptation and creative development of the achievements of direct cinema. He was able to forge his negative attitude to Hollywood and what

[19] The same plot was subsequently employed by Luc Besson in *Léon* (1994). A remake of *Gloria* was also made, featuring Sharon Stone as the lead (1999, directed by Sidney Lumet).

[20] Cook, *Lost Illusions*, 119–32.

[21] Ibid., 130.

it represented in the 1950s and 1960s into a strategy which allowed him to construct his own, inimitable form of expression.[22]

Contemporary viewers are accustomed to a broad array of forms. In particular circumstances—for instance, when watching documentaries shot with a candid camera—they are even ready to regard the image from an industrial camera as a form of artistic expression which employs a particular style. They possess an astounding ease in switching from the aesthetically refined images of Bergman to video clips and commercials (very often no less aesthetically refined). They are surprised to hear that the viewers of the famous Grand Café screening fled at the sight of the train approaching the La Ciotat station, that they hid under their seats when the gangster in *The Great Train Robbery* (1903, directed by Edwin S. Porter) pointed his gun at the camera, or that, when their favorite actor was shown in *plan américain* (from the knees upwards), they demanded to be shown a man in full rather than with his legs cut off when they have paid dearly for their tickets. John Cassavetes entered the history of cinema very early; this is why he had to wait until the 1980s for a positive reception of his work, and even longer for his followers to appear.

Works Cited

Carney, Raymond, ed. *Cassavetes on Cassavetes*. London and New York: Faber and Faber, 2001.

—. *American Dreaming: The Films of John Cassavetes and the American Experience*. Berkeley, Los Angeles, and London: University of California Press, 1985.

—. *The Films of John Cassavetes: Pragmatism, Modernism, and the Movies.* Cambridge, New York, and Melbourne: Cambridge University Press, 1994.

Cassavetes, John. "What's Wrong with Hollywood?" *Film Culture* 19 (April 1959): 22–23.

Cook, David. *Lost Illusions: American Cinema in the Shadow of Watergate and Vietnam, 1970–1979*. Berkeley, Los Angeles, and London: University of California Press, 2002.

Durys, Elżbieta. "Improwizacja w *Cieniach* Johna Cassavetesa." In *Studia Filmoznawcze*, edited by Sławomir Bobowski, 23:49–60. Wrocław: Wydawictwo Uniwersytetu Wrocławskiego, 2002.

---

[22] Appreciation of his work, however, is growing steadily, especially among the aforementioned historians and critics writing about American independent cinema.

Fine, Marshall. *Accidental Genius: How John Cassavetes Invented the American Independent Film*. New York: Hyperion, 2005.

Hendrykowski, Marek. *Słownik terminów filmowych*. Poznań: Ars Nova, 1994.

King, Geoff. *American Independent Cinema*. London and New York: I. B. Tauris, 2005.

Kouvaros, George. *Where Does It Happen? John Cassavetes and Cinema at the Breaking Point*. Minneapolis, MN: University of Minnesota Press, 2004.

Mętrak, Krzysztof. "Miazga Cassavetesa." *Kino*, no. 3 (1979): 62–63.

Przylipiak, Mirosław. "Kino bezpośrednie." Pts. 1 and 2. *Kino*, no. 1 (2001): 45–48; no. 2 (2001): 50–53.

—. *Kino bezpośrednie, 1960–1963*. Gdańsk: słowo/obraz terytoria, 2007.

—. *Poetyka kina dokumentalnego*. 2nd ed. Gdańsk: Wydawnictwo Uniwersytetu Gdańskiego; Słupsk: Wydawnictwo Pomorskiej Akademii Pedagogicznej, 2004.

Tzioumakis, Yannis. *American Independent Cinema: An Introduction*. New Brunswick, NJ: Rutgers University Press, 2006.

CONTRIBUTORS

Raffaele Asquer is a Ph.D. student at the Department of Political Science, University of California, Los Angeles. Attending both Scuola Normale Superiore di Pisa and University of Pisa (Italy), he studied Contemporary History with a focus on American History, graduating in 2007 with a research thesis on the public response to Franklin D. Roosevelt's fireside chats. Admitted to UCLA with a Shapiro Fellowship, he is currently enrolled as a first-year graduate student in the Department of Political Science. His work focuses on American political institutions, and particularly on the historical development of the modern presidency.

Tomasz Basiuk teaches at the University of Warsaw American Studies Center, which he has directed since 2005. His degrees are from the University of Warsaw (M.A. 1989, Ph.D. 1997). He was recipient of academic scholarships in the United States of America, the United Kingdom, and Germany. In 2004–2005 he was Fulbright Senior Scholar at the CUNY Graduate Center in New York City. He wrote a book-length study of William Gaddis: *Wielki Gaddis: realista postmodernistyczny* (2003), and co-edited three volumes of essays on queer studies: *Odmiany odmieńca/A Queer Mixture* (2002), *Parametry pożądania* (2006), and *Out Here* (U.K. 2006). He co-edits the scholarly e-journal *InterAlia* (www.interalia.org.pl). His current project is on gay men's life writing.

Hanna Boguta-Marchel is a fifth year doctoral student at the American Literature Section of the Institute of English Studies at the University of Warsaw. In 2003 she received an M.A. from the Institute of English Studies (her master's thesis on Puritan ideology was written under the supervision of Professor Agata Preis-Smith) and in 2004 an M.A. from the Faculty of Psychology, both at the University of Warsaw. Her doctoral thesis in progress concerns the issue of evil and violence as dealt with in the novels of Cormac McCarthy and read in the light of various theological and philosophical theories of evil. She teaches English and translation at the University of Warsaw and the Warsaw School of Social Psychology.

Kate Delaney teaches literature and writing at the Massachusetts Institute of Technology. She is a member of the editorial board of *American Studies* and coordinator of the international initiative of the American Studies Association. Her essays appear in various US and European journals and compilations. She is working on a study of culture in the post-Cold War era.

Elżbieta Durys is an Assistant Professor at the Department of Transatlantic and Media Studies, University of Łódź (Poland). Her research interests focus on film studies, anthropology of culture, communication and gender studies. She is the author of numerous articles on film that have appeared in journals as well as in anthologies. She is the co-editor of *Wizerunki kobiet i mężczyzn w kulturze* (Representing Gender in Cultures), 2005; *Kino amerykańskie: Dzieła* (American Cinema: The Works), 2006; *Kino amerykańskie: Twórcy* (American Cinema: The Authors), 2007. Her book devoted to John Cassavetes' oeuvre is to be published in 2009.

Julia Fiedorczuk, Ph.D., teaches American literature at the Institute of English Studies, University of Warsaw, Poland. Her research interests include American poetry, experimental fiction, literary theory, psychoanalysis and ecocriticism. Member of ASLE (Association for the Study of Literature and the Environment). She is a poet with four books of her poetry published. She has also translated into Polish the writings of John Ashbery, Wallace Stevens, Laura (Riding) Jackson, and other American authors.

Małgorzata Gajda-Łaszewska is an Assistant Professor at the American Studies Center, University of Warsaw, Poland. She deals with issues of representation and formulation of media messages. She teaches classes on American media, advertising and consumer culture. Her leading interest focuses on the issues of representation of war in American media and culture. Her Ph.D. dissertation, defended in 2003, was devoted to researching how the presentation of the Gulf War of 1991 in the media fit and reflected the mainstream culture of modern America.

Dorota Golańska holds a Ph.D. in Humanities and an M.A. in International Relations. She is an Assistant Professor at the Department of Transatlantic and Media Studies and affiliate researcher at the Women's Studies Center, University of Łódź. Her research activities fall into the field of interdisciplinary gender studies. She is interested in feminist theories of visual culture, criticism of representation as well as in different intersections of culture and politics. She has recently worked with such diverse concepts as intimate citizenship and nationalism on the one hand, and visualization technologies, critique of representation, and affirmative turn on the other. She has also published in all these fields.

Ewa Grzeszczyk, Ph.D., was a sociologist of culture with academic background in Film Studies as well as Art History. The author of a book *Success: American Models, Polish Reality*, in her last research project she focused on the spheres of American popular culture, consumption, religiosity and the Americanization of Polish popular culture. Dr Grzeszczyk worked at the American Studies Center, University of Warsaw, teaching courses on American society and popular culture. Earlier she also worked as a researcher at the Institute of Philosophy and Sociology of the Polish Academy of Sciences and a visiting scholar at the University of Notre Dame. She died May 22, 2009.

David A. Jones is Professor of Law, Management, Policy, and Strategic Studies at the University of Warsaw, Poland, and director of its United States Law Program. He offers its Lecture on United States Foreign Policy, Lecture on European Union Single Market, and law courses at its American Studies Center and Institute of International Relations. His forthcoming book, *The Eagle and the Dragon*, compares 21st century China to 19th and 20th century America. He is the author of many books, and was academic dean responsible for the Schools of Business, Law, Media, and Public Policy at Concordia International University, Tallinn, Estonia. Professor Jones has held an American Federation of Television and Radio Artists (AFTRA) union card, and hosted radio talk shows in the United States.

Izabella Kimak, M.A. is a doctoral student at the Department of American Literature and Culture, Maria Curie-Skłodowska University in Lublin, Poland. She is currently working on her Ph.D. dissertation devoted to the representation of female sexuality and female body in fiction by South Asian American women authors. Her research interests include contemporary American literature and culture, literatures of ethnic minorities, and gender studies.

Marta Koval is an Associate Professor at Lviv Regional Institute of Public Administration, where she chairs the Department of Applied Linguistics. She received her Ph.D. from the Institute of Literature of the National Academy of Sciences of Ukraine in 2000. Her special academic interests include postwar American fiction and transformations of the historical novel in the late 20th/early 21st century. In 2000, she published her book *Play in the Novel, Playing Novel: On John Barth's Fiction.* She is the author of many essays on postwar American fiction. In 2004–2005 she was a Senior Fulbright Fellow at Michigan State University.

Grzegorz Kość holds a doctorate in humanities and teaches American literature and culture at the English Institute of the University of Łódź and at the American Studies Center of Warsaw University. He is interested in twentieth-century American poetry, public poetry in particular, political philosophy and photography. He is an author of *Robert Lowell: Uncomfortable Epigone of the Grands Maîtres* (Frankfurt: Peter Lang, 2005). Presently he is working on a book on Robert Frost as a national poet.

Anna Krawczyk-Łaskarzewska graduated from the University of Warsaw (Ph.D. in American Literature). Since 1995 she has been on the faculty of the English Department at the University of Warmia and Mazury in Olsztyn. She has published articles and presented papers on the prose of Ambrose Bierce and H. P. Lovecraft, the history and challenges of the discipline of American Studies, the theory and practice of literary translation, as well as selected pop culture phenomena. Her current research focuses on British cinema and modern-day film adaptations of the so-called literary canon.

Krzysztof Majer is involved both in American and Canadian studies. He holds a Ph.D. degree from the University of Łódź; the title of his thesis is *The Picaro Messiah and the Unworthy Scribe: A Pattern of Obsession in Mordecai Richler's Later Novels*. His academic interests include North American Jewish writing and post-war fiction. He has published articles on Mordecai Richler, Mark Anthony Jarman and John Barth. He is a member of BACS (British Association for Canadian Studies), PACS (Polish Association for Canadian Studies) and PAAS (Polish Association for American Studies). He is currently employed at Adam Mickiewicz University, Kalisz, Poland.

Mirosław Aleksander Miernik defended his M.A. thesis (entitled *The Extent to which Selected Youth Subcultural Movements in 1969–1983 in Britain Were Reactive to the Social, Economic and Political Developments of the Period*) in 2005 at the Institute of English Studies, University of Warsaw. His academic interests include semiotics, modern American literature, science fiction literature as well as literary theory. He is currently writing his Ph.D. thesis.

Alicja Piechucka is an Assistant Professor at the Department of American Literature and Culture, Institute of English Studies, University of Łódź. She received her Ph.D. from the University of Łódź in 2006. She specializes in American literature and culture. Her academic interests include American modernist poetry, French literature and visual arts. Her research focuses on interdisciplinary studies, in particular comparative literature. Among the courses she teaches are US literature and culture as well as American painting.

Noel Polk is Professor Emeritus of English and editor of *The Mississippi Quarterly* at Mississippi State University. He has published and lectured on Eudora Welty and William Faulkner in the USA, Europe, Japan, and the former Soviet Union. He has edited the works of William Faulkner and Robert Penn Warren. Other publications include *Children of the Dark House: Text and Context in Faulkner*; *Eudora Welty: A Bibliography of Her Work*; *Outside the Southern Myth* as well as *Faulkner and Welty and the Southern Literary Tradition.*

Richard Profozich is a Professor of American Literature and Culture at the University of Łódź. His research interests are in American Studies, including topics in media and journalism, American style democracy, foreign policy, and distance learning.

Andrew Ross is Professor and Chair of the Department of Social and Cultural Analysis at New York University. He is the author of several books, including *Nice Work if You Can Get It: Life and Labor in Precarious Times* (2009); *Fast Boat to China: Corporate Flight and the Consequences of Free Trade; Lessons from Shanghai* (2006); *Low Pay, High Profile: The Global Push for Fair Labor* (2004); *No-Collar: The Humane Workplace and Its Hidden Costs* (2004) and *The Celebration Chronicles: Life, Liberty and the Pursuit of Property Value in Disney's New Town* (2000). He has also edited several books, including *No Sweat: Fashion, Free Trade, and the Rights of Garment Workers*; *Anti-Americanism* (1997) and *The University Against Itself: The NYU Strike and the Future of the Academic Workplace* (2008).

Aleksandra M. Różalska is an Assistant Professor at the Department of Transatlantic and Media Studies, University of Łódź, Poland. In 2006 she defended her Ph.D. dissertation titled *Representations of Ethnic Minorities in Contemporary American Television*. Since 2002 she has been an affiliate of the Women's Studies Centre, University of Łódź. In 2005 she was a Fulbright fellow at the American University, Washington, DC. Apart from studies on television, multiculturalism and diversity in the US and in Europe, her research interests include: intersections of gender, ethnicity, and class in the media; relationships between the media and politics as well as audience studies. She has published on images of minorities and women in film and television as well as on theories of spectatorship.

Radosław Rybkowski defended a dissertation on the strategies of promoting actresses' career in 19th century American theater in March 2000. Later that year he started his academic work at the Institute of Public Affairs of the Jagiellonian University, where he taught courses on cultural policy and basics of arts administration. Since 2001 he has been working at the Center for American Studies, teaching courses such as the American Civilization, History of American Theater and American Higher Education Policy. During his academic career he has been awarded fellowships from Sasakawa Young Leaders' Fund, Salzburg Seminar, and Polish-American Fulbright Commission and focused in his research on the interrelation between public policy and culture. In 2005 and 2008 he taught at the Department of Political Science, University of Rochester. At present he is focused on comparative research on the American higher education policy and the development of American culture.

Maciej Turek holds degrees in Political Science (B.A., 2003, Pułtusk Academy of Humanities, Poland) and Culture Studies (M.A., 2006, Jagiellonian University, Poland), minoring in American Studies in both programs. Since 2006, he has been enrolled in the Ph.D. program at the Institute of American Studies and Polish Diaspora, Jagiellonian University, where he devotes his research to the problems of the institution of the Vice Presidency of the United States. His research interests also include the electoral process in the United States, state and local government, U.S. in the 1970s, and relations between federal U.S. political institutions.

Joanna Waluk holds her M.A. degree from the American Studies Center, Warsaw University. She is a lecturer of English at the University of Bialystok, Poland. Her interests include twentieth century American literature and culture as well as methodology of teaching English as a foreign language. She is a member of the Polish Association for American Studies.

Matthew Wilson is Professor of Humanities and English in the School of Humanities at Penn State Harrisburg. He has published *Whiteness in the Novels of Charles W. Chesnutt*, edited three of Chesnutt's manuscript novels, and a selection of letters between Chesnutt and George Washington Cable. In addition, he has published a number of essays on contemporary fiction—on John Edgar Wideman, David Bradley, Philip Roth, and John Updike. In the academic year 2007–2008, he was Fulbright Distinguished Chair in American Literature and Culture in the Department of American Literature and Culture at the University of Łódź, Poland.

Marek M. Wojtaszek is currently completing his Ph.D. studies in the Transatlantic Studies and Mass Media Department at the University of Łódź, Poland. His research reflects his critical engagement with poststructuralist feminist philosophies, visual studies, and Gilles Deleuze's philosophy. He has published a range of articles on contemporary American film, fiction, and visual arts.

Beata Zawadka is an Assistant Professor in the Chair of English Philology at the Szczecin University, Poland. She earned her Ph.D. from The University of Łódź. Her doctoral dissertation was on the work of Peter Taylor, particularly the identity of his women characters. She specializes in southern literary studies. Her research focuses on literature by contemporary southern women writers such as Ellen Gilchrist, Jill McCorkle, and Jayne Anne Philips. She is also interested in Arab-American women writers (Diana Abu-Jaber, Pauline Kaldas, Laila Halaby, Mohja Kahf), popular culture and American cinema (Hollywood genres). She is a member of the EAAS (PAAS and ASAT) and of the Southern Studies Forum of the EAAS. Dr Zawadka has presented papers at conferences in the European Union and has also published—in Poland and abroad—on topics linked to her fields of research.

INDEX